THE LITTLE BOOK of MUSHROOMS

An ILLUSTRATED GUIDE to
the EXTRAORDINARY POWER of
∽ MUSHROOMS ∽

Alex Dorr
Illustrated by Sara Richard

ADAMS MEDIA
New York London Toronto Sydney New Delhi

Adamsmedia

Adams Media
An Imprint of Simon & Schuster, Inc.
100 Technology Center Drive
Stoughton, Massachusetts 02072

First Adams Media hardcover edition May 2023

ADAMS MEDIA and colophon are trademarks of
Simon & Schuster.

For information about special discounts for bulk purchases,
please contact Simon & Schuster Special Sales at
1-866-506-1949 or business@simonandschuster.com.

The Simon & Schuster Speakers Bureau can bring
authors to your live event. For more information or to
book an event contact the Simon & Schuster Speakers
Bureau at 1-866-248-3049 or visit our website at
www.simonspeakers.com.

Interior design and hand lettering by Priscilla Yuen
Illustrations by Sara Richard
Interior images © 123RF/Amphawan Chanunpha, archnoi1,
dimarik16, drakonova, KATSUMI MUROUCHI, magenta10,
Olga Zakharova, pauldesign, pgmart, Roman Boiko,
shumo4ka, Vasya Kobelev; Simon & Schuster, Inc.

Manufactured in the United States of America

10   9   8   7   6   5   4

Library of Congress Cataloging-in-Publication Data
has been applied for.

ISBN 978-1-5072-1959-1
ISBN 978-1-5072-1960-7 (ebook)

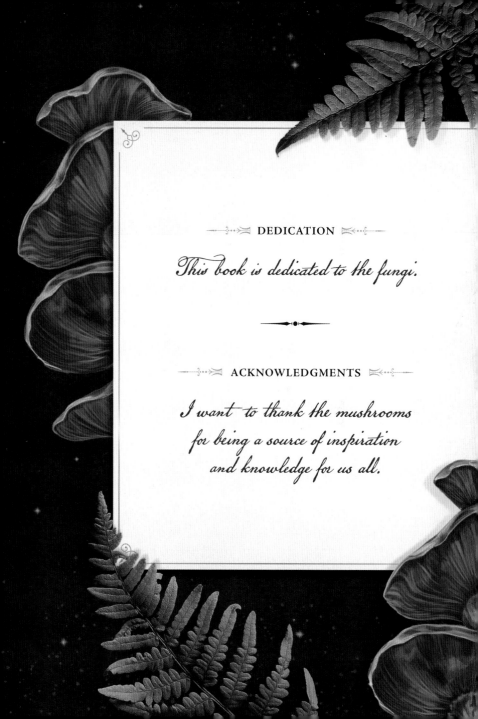

⟶⟶⟶⟶⟶ DEDICATION ⟶⟶⟶⟶⟶

This book is dedicated to the fungi.

⟶⟶⟶⟶⟶ ACKNOWLEDGMENTS ⟶⟶⟶⟶⟶

*I want to thank the mushrooms
for being a source of inspiration
and knowledge for us all.*

Contents

Index **253**

Introduction

WELCOME TO THE MYSTERIOUS and often misunderstood world of mushrooms. These organisms are a wonder, in that they aren't quite plants, animals, or bacteria. They are something different to celebrate and explore.

Mushrooms are recyclers, bridges between life and death, healers, feeders, teachers, killers, and so much more. Some can thrive in radioactive environments. Others can hibernate underwater for millions of years. And still, others may brave the extreme cold of Antarctica, or the desolation of space. And here, throughout *The Little Book of Mushrooms*, you'll find seventy-five beautifully illustrated mushroom profiles that shine a light on this mysterious world, allowing you to revel in the remarkable properties of these unique, beautiful, and scientifically astounding fungi. Each profile—from the delicious Aborted Entoloma to the compelling Zombie-Ant Fungus—will show you:

+ Where each of these mushrooms lives.
+ The mushroom's preferred habitat and growing conditions.
+ Any physical attributes that distinguish each fungus.
+ What, if anything, each species of mushroom is used for.

You'll also learn more about the cultural history and myriad uses of these diverse fungi, come to appreciate the complex process that takes them from spore to mushroom, and discover just how varied the kingdom of these seemingly simple organisms really is.

So whether you're specifically a mushroom enthusiast or just a casual observer of the natural world in general, *The Little Book of Mushrooms* is guaranteed to help you celebrate the cryptic world of fungi.

Let's begin.

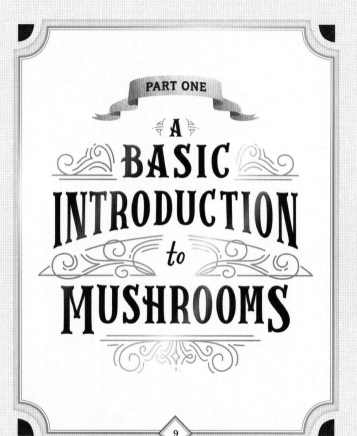

PART ONE

A BASIC INTRODUCTION to MUSHROOMS

MUSHROOMS ARE AN INTEGRAL PART OF LIFE ON EARTH and have a long history of worldwide relevance. They help save lives, create forest ecosystems, store carbon, make new materials, build soil, and feed people, animals, plants, and all of life as we know it. This first part of the book distinguishes the difference between mushrooms and fungi, but it will also:

+ Break down the life cycle of a mushroom.
+ Describe the different types of mushrooms.
+ Provide a list of mycology-related terms.
+ Explain key components of the mushroom entries in the second part of this book.

Before delving into a few scientific details about mushrooms and some related terminology, we will take a brief look at the historical significance of mushrooms and their enduring popularity.

Why Mushrooms?

Mushrooms have been the cornerstone of many cultural ceremonies and rituals throughout history. Herbalists and others have been using mushrooms for centuries to support human health. Psilocybin Mushrooms, known for their psychoactive properties, may have been used as far back as twelve thousand years ago in spiritual practices. Indigenous peoples of Central America featured these mushrooms in their religious ceremonies and art and called them "flesh of the gods." Aztecs and Mayans, as well as other groups worldwide, have used Psilocybin for healing and to connect with the Divine.

Many kinds of mushrooms have been used traditionally in herbal practices. More than two thousand years ago, the ancient Greek physician Hippocrates mentioned the Agarikon Mushroom's cure-all capabilities, including against potentially fatal viral diseases. Traditional Chinese herbalists have prescribed mushroom blends to support overall health and wellness. Some species are believed to help with supporting cardiovascular health, some to support energy levels, and others to promote optimal immune system function.

The use, appreciation, and wonder of the mushroom have not lessened over time. Centuries later, the mushroom is still at the forefront of both popular and scientific culture. These organisms have a vast number of modern-day applications. For the DIY or art enthusiast, multiple species of mushrooms can be used to make paper, and as an added bonus, other species can be used to create pigment for uses such as ink. Other species, such as the Tiger Sawgill, are used for mycoremediation, or removing pollutants from the environment. The Sawgill may be used for site-runoff, while different mushrooms in the mycoremediation realm may focus on removing mercury or other undesirable things from the soil. To this day, many mushrooms have not revealed their true functional or scientific potential, though studies are constantly underway. Now that you know why these organisms are relevant and worth celebrating, the question remains:

What makes a mushroom a mushroom?

What Are Mushrooms?

 Mushrooms are complex organisms with unique characteristics that distinguish them from plants. Instead of seeds, mushrooms have microscopic spores that spread in a variety of ways. Some spores are dispersed via animals' willing (or unwilling) cooperation, while some are spread by a passing breeze. When these spores land on a life-sustaining material, they will grow mycelium (microscopic roots).

Mushrooms typically grow in easily accessible organic material like leaves or trees, but you can find them in any number of places worldwide, including Antarctica. Mushrooms also thrive in otherwise inhospitable living conditions, like in the nuclear power plant at Chernobyl, one of the most radioactive sites on the planet. They are truly different from any other organism. However, people often mistake mushrooms as being the same exact thing as everything in their larger classification, or kingdom, of Fungi. While the organisms in this kingdom are closely related—all mushroom species are, in fact, fungi—the word *fungi* does not solely refer to mushrooms.

Mushrooms versus Fungi

The Fungi kingdom includes much more than just mushrooms. Fungi consist of both multicellular organisms and unicellular organisms; in addition to mushrooms, they may take the form of any of the following:

- molds
- yeasts
- mildews
- rust plant diseases
- smut plant diseases

All types of fungi share some key characteristics. They cannot produce their own food (unlike plants, which photosynthesize);

many make efficient decomposers; and they reproduce through the release of spores. Certain types of fungi, if given proper circumstances, may result in the fruiting body of a mushroom. However, only certain types of fungi do this, and only if the appropriate environment presents itself. For example, some mushroom species grow only on certain types of trees; without those trees, mushroom growth will not occur. Think of mushroom reproduction like apples on an apple tree. Like apples, which are fruiting bodies containing seeds of the apple tree, mushrooms are the fruiting bodies of a fungus. Instead of seeds, they contain spores. Mushrooms help distribute spores when eaten or otherwise spread, just as apples distribute seeds. When the spores spread, a mushroom's DNA spreads, and the species can survive.

The Life Cycle of Mushrooms

Mushrooms have interesting, intricate growing patterns. The mushroom's life cycle is typically broken up into four parts:

✦ mushrooms   ✦ spores

✦ hyphae   ✦ mycelium

Remember: All mushrooms are different, and this is a broad generalization of the most common life span.

Mushrooms

The types of mushrooms come in many different structures, colors, shapes, sizes, and mechanisms of reproduction. Some are edible, some are poisonous, some glow in the dark, some are squishy. Others are rock-hard, or the size of a small child, or as small as your pinky nail. For the sake of illustration, let's look at the common cap-and-stem mushrooms you see at the grocery store.

These and other mushrooms function as a fruiting body, which spreads its spores so that its gene pool continues. The way that this reproductive structure represents itself may vary tremendously based on the species. A mushroom may have gills or pores to make spreading its spores easier. Every mushroom is unique, but their purpose remains the same: to spread spores.

Spores

Once the spores are released into the environment, they are carried by animals, water, or wind to new places. Some mushrooms can release millions of spores every day, giving a higher probability of success for the survival of the organism. Other species require such specific conditions to grow that it's miraculous if a few new mushrooms sprout. Each spore released contains 50 percent of the DNA required to reproduce. In order to find that second 50 percent, the spore will transform into a new structure: a hypha.

Hyphae

Once spores land in an appropriate growing medium, such as soil, wood, an insect, or dung, they will germinate and become hyphae. Hyphae are one-celled filamentous probes that push through the organic material in search of a suitable partner. Most living organisms, including primitive fungi, have two mating types. Generally, this means that only one mating type (commonly referred to as "male") may reproduce with one other specific mating type (a "female"). Fungi have a *much* wider net of potential reproductive partners. Some fungi have up to 23,328 mating types, meaning each hypha or spore has thousands of different mating options available to them. Once just one of these possible combinations happens, the organism moves to the next stage of life.

Mycelium

After the hyphae fuse, they create a massive web-like structure, known to grow as large as 2,384 acres, called mycelium. Mycelium's role in the ecosystem is to excrete enzymes which decompose everything in the vicinity and then use the broken-down nutrients to grow. When the mycelium encounters a threat, knows it's running out of food, or the outside environment is just right, it will produce structures called mushrooms. The life cycle then begins again.

Kinds of Mushrooms

Mushrooms come in many shapes, sizes, colors, and patterns. The more you go out in the woods and pay attention to the life around you, the more you will see and be astonished by the complex biodiversity of the fungal world. Mushrooms are just one small piece of a much wider biological web of life. Fungi are the great connectors and disassemblers of our ecosystems, and without fungi, life as we know it would cease to exist. Fungi as a giant umbrella can be broken up into two main ecological roles: decomposers and mutualistic fungi. The different types may vary somewhat in attributes of their life cycles. In relation to its environment, each type has different benefits or potential drawbacks. Decomposers capitalize on harming, or taking advantage of, another organism's resources, whether alive or dead. Mutualistic mushrooms (and fungi) do the opposite: They create symbiotic relationships, meaning that they interact with other species so both organisms gain something from their relationship. These two types of fungi each have a few separate subtypes, making the diversity of the mushroom world that much greater.

Decomposers

When a tree falls in the forest, mushrooms (or various other types of fungi) help decompose it. How do they do this? When humans eat, we have digestive enzymes that help break down our food. Decomposer fungi, on the other hand, excrete these digestive enzymes outside of themselves to help break down their food (in this case the log) externally. Then they absorb the nutrients after the fallen tree, for example, has begun to decompose. They are the great recyclers in our environment, turning waste into soil so our ecosystems can thrive.

PARASITIC

Many mushrooms or fungi function as parasites, meaning they leech off the resources of another living thing to grow and gain nutrients. Mushrooms may feed on other fungi, animals, and plants. Regardless of how they go about it, these fungi dominate their host, and then get to their main objective: reproducing.

FUNGI Some mushrooms or fungi act as parasites of other fungi. They will take over the victim fungus's body and utilize the decomposing body to make nutrients for itself. Examples of these parasites include the famous edible Lobster Mushroom or the Aborted Entoloma.

. . . .

ANIMALS Some fungi attack and grow in, or on, animals. Many types of fungi attack the human body. Some of the most well-known fungi that may use the human body as a host include ringworm, athlete's foot, toenail fungus, yeast infections, and jock itch. Although mushrooms are not among the types of fungi that might live on the human body, they do attack insects. Mushrooms that parasitize insects are referred to as entomopathogenic fungi; some of these species are in the *Cordyceps* genus. Cordyceps mushrooms turn insects into zombies, making them climb up to a high place in a tree and then sprout mushrooms out of their head. Fungi can also infect a wide range of other living creatures, including mammals, reptiles, microscopic animals, and worms.

SAPROPHYTIC

Saprophytic fungi are the only type of decomposers that will focus on already dead or decomposing organic matter, like fallen logs, dung, or dead animals. Think of the mold that covers a rotting orange on your kitchen counter if you forget about it. This would be a prime example of a saprophytic fungi breaking down the orange into nutrients for the soil. If this were on the forest floor, the organic breakdown would then fuel the soil. Usually, the types of saprophytic fungi are divided into one of two main categories: white rot fungi or brown rot fungi (with more subcategories as well). White rot fungi refer to fungi that mostly target the lignin and leave behind the white-colored cellulose and hemicellulose, which are all just fancy words for different structural components in plants. Brown rot fungi primarily target the cellulose and hemicellulose, leaving behind the brown-colored lignin (aka: different structures that make up plants). Some examples of wood-decaying mushrooms are Turkey Tail, Chicken of the Woods, and Shiitake.

Mutualistic

Mushrooms can be great teachers when it comes to creating symbiosis with other living organisms. From insects, to humans, to plants, to bacteria, mushrooms (and fungi as a whole) connect with the world around them in the most interesting ways.

Decomposer fungi rely on other organisms to provide them nutrients by harming them (in the case of parasitic mushrooms) or taking advantage of the already dying or dead organic matter (saprophytic). Mutualistic fungi are the opposite. With mutualism, two organisms (the mushroom or fungi and its mutualistic partner) are affected, but both in positive ways. They both benefit from the relationship.

MYCORRHIZAL

About 90 percent of all terrestrial plants have mycorrhizal connections with fungi, meaning they have a symbiotic relationship at the roots. The fungi trades nutrients with the plant and in return the plant gives the fungi certain sugars as food. The fungi also protect the plants from invaders and help store water. Porcini Mushrooms, one of the world's most sought-after gourmet mushrooms, are one example of mushrooms that establish a mycorrhizal relationship with a plant. The mycelium connects with the roots of deciduous and coniferous trees to trade nutrients. Mushrooms then grow aboveground for spore dispersal.

ENDOPHYTIC FUNGI

All plants also have endophytic relationships in which fungi live inside the cell walls and leaves of the plant, helping the plant and receiving help in return. Although no mushrooms grow from this type of relationship, scientists are learning that these endophytic fungi help influence the chemicals that the plant produces. For example, the taste and smell of spearmint is highly influenced by a fungus living in the cell walls of the plant.

LICHEN

Lichen forms when cyanobacteria live in the filaments of multiple fungi. You can see these flat and frilly bright orange to blue to green to yellow to red organic structures on old park benches, tree bark, gravestones, rocks, fences, and more. The fungi provide structure and protection for the cyanobacteria while the cyanobacteria gain nutrients from photosynthesis. They work together in symbiosis and were the first terrestrial organisms to form this type of symbiotic relationship around seven hundred million years ago. It is thought that lichen are the early ancestors to present-day mushrooms and plants.

BACTERIA

Many kinds of bacteria can live on and inside fungi, a relationship that has been influencing genetics and evolution for millions of years. Among other benefits, the bacteria can protect fungi from parasitic nematodes and help regulate reproduction of the fungi.

ANIMALS

Many insects, including leaf-cutter ants, form symbiotic relationships with fungi. Leaf-cutter ants bring leaves back to their nest and chew them into a paste. The queen leaf-cutter ant then takes a chunk of a mushroom and inoculates the leaf litter mash and from it mycelium grows. The more the ants continue to feed this mycelium with more leaf litter mash, the more mycelium grows from it. The ants use the mycelium that grows on the leaf litter as their sole food source. The author, Alex Dorr, researched leaf-cutter ants in Ecuador in 2015 and found that a single nest can bring in up to 1,500 pounds of leaf litter per year to transform it into fungal mycelium for food.

Squirrels and deer have been known to be mycophagous, meaning they eat mushrooms and other fungi. People use this to their advantage when they search for truffles, which are a type of fungus that grows underground. The truffle hunters keep their ears open for squirrel noises, indicating the squirrels are trying to protect their truffle stashes, and the hunters dig where squirrels are making the most noise. Birds also have been known to bring mushrooms back to their nests to decorate them, which attracts mates and helps spread the spores when a mate visits the nest.

We as humans also have a symbiotic relationship with fungi. When we forage and cultivate mushrooms, such as Oyster Mushrooms, we are helping them reproduce and grow while we get food and functional health benefits in return.

Terms and Phrases to Know

 Much of the terminology around mushrooms may feel or sound academic in nature. This section is provided so that you can easily refresh your memory related to any unfamiliar or forgotten terminology as you read through the book.

► adaptogen
An herb or mushroom that helps the body adapt to occasional stress.

► ascomycota
A phylum of the kingdom Fungi that includes brewer's and baker's yeast, penicillium, truffles, Morels, Dead Man's Fingers, and more.

► cap
Also called the pileus, the structure that houses the spore-bearing surface of a mushroom.

► conk
A hard, woody, and shelf-like mushroom structure that grows from the side of trees.

► endophytic
Fungi that live in the cell walls of plants.

► entomopathogenic fungi
Fungi that attack insects or other arthropods.

► enzymes
Proteins that help break apart food.

functional

Describes types of mushrooms that are considered to have certain health and wellness benefits. The most common benefits of mushrooms are to support energy levels, immune system function, memory and cognitive function, relaxation, and more. Each mushroom is different and will offer different benefits to the body.

fungi (singular: fungus)

Organisms that are part of the kingdom Fungi; includes mushrooms as well as molds, yeasts, and plant diseases known as rusts and smuts.

gasterothecium

A sphere shape that a puffball mushroom makes.

genus (plural: genera)

Part of the taxonomic umbrella that classifies all living organisms. Genus is above the species level. Human beings, *Homo sapiens*, are in the genus *Homo* and species *sapiens*.

gills

Found on the underside of some mushroom caps and housing the spores. Some mushrooms have gills, some have teeth, and others have pores. Whatever the structure, they all contain the spores used to carry the mushroom's DNA.

hyphae

A branching single-celled filamentous structure that emerges from a fungal spore and makes up mycelium.

indusium

A skirt-shaped organic net that comes from the cap of some stinkhorn mushrooms.

kingdom

Part of the taxonomic umbrella that classifies all living organisms. There are six kingdoms (Animalia, Plantae, Fungi, Protista, Archaea/Archaebacteria, and Bacteria/Eubacteria). Some countries combine Archaea/Archaebacteria and Bacteria/Eubacteria into one kingdom and call it Monera, therefore identifying only five kingdoms of life.

mushroom

The fruiting structure of a fungus, similar to an apple on an apple tree.

mycelium

Similar to the roots of a plant, these structures secrete enzymes to break down their food and absorb nutrients. Only some species that have mycelium will produce mushrooms.

mycoremediation

Removing pollutants from the environment with fungi.

mycorrhizal
A symbiosis between plant roots and fungi.

nonpolar compounds
Compounds in fungi that can be extracted with alcohol.

nutrient cycling
Energy or matter exchanged between living and nonliving portions of the environment.

parasitic
An organism that survives by living on or in another species.

peridium
An organic layer protecting spores in fungus.

polar compounds
Compounds in fungi that can be extracted with water.

polypore
A group of mushrooms with pores or tubes on the bottom of the cap where spores come out.

pores
Found on the underside of some mushroom caps and housing the spores.

psychoactive
Also referred to as psychedelic, mushrooms of this type will get you high.

saprophytic
Mushrooms that decompose dead or dying organic matter for food.

sclerotium
An underground mass of hardened mycelium that the fungi use as long-term food storage.

species
Part of the taxonomic umbrella that classifies all living organisms. The main groups, organized in a hierarchy from highest to lowest, are: Kingdom, then Phylum, Class, Order, Family, Genus, and finally, Species. Species is the lowest differentiating classification. For example, human beings, *Homo sapiens*, are in the genus *Homo* and species *sapiens*.

spores
The equivalent of seeds for a mushroom, which carry genetic material.

stem
Also called a stipe, its purpose is to elevate the cap for more spore dispersal.

substrate
The organic material that mushrooms fruit off of, such as straw or sawdust.

teeth
Found on the underside of some mushroom caps and housing the spores.

More about Mushrooms

Part 2 of this book is full of beautiful illustrations and detailed information about seventy-five different kinds of mushrooms. Before we get there, let's take a moment to help you understand the information in each mushroom entry, including the scientific and common names for mushrooms, where mushrooms grow, what they look like, and for what purpose they are primarily used. To start, we will decode the difference between common names and scientific names.

Scientific Name versus Common Name

Each mushroom in this book has at least two names: a scientific name, and one or more common names. A mushroom's scientific name is the biological Latin name that describes the genus and species that the mushroom falls into. When learning about mushrooms, it's always best to learn the scientific name of the mushroom because no matter where you travel in the world, the Latin name is universal. There can sometimes be dozens of different common names for the same mushroom, but there should always be one Latin name. For example, a mushroom with the scientific name *Boletus edulis* might be called by any number of common names, such as Porcini, Penny Bun, Porcino, Hog Mushroom, King Bolete, Noble Mushroom, Belyy Grib, or Cep. "Common names," as the term suggests, are what the mushrooms are most commonly known by. Certain common names are umbrella terms that refer to multiple species. An example would be Destroying Angel Mushrooms; this name is used for four related mushroom species (*Amanita bisporigera, Amanita ocreata, Amanita virosa*, and *Amanita verna*), which are all listed in the subtitle of the entry. Thus, using the scientific name is assuredly more accurate.

Geographic Location

The geographic location given in each entry tells you where in the world the mushroom or fungus is most likely to be found: the continents, countries, regions, or habitats. Remember, these geographic locations are only what scientists have discovered to date, and these mushrooms could show up in more places as research continues.

Growing Location

The growing location describes the ecological niche a mushroom inhabits, as well as what that mushroom needs to grow. Some mushrooms grow on (and decompose) wood, others form a symbiotic relationship with plant roots or decompose dung, and some parasitize other mushrooms or insects. Some mushrooms love tropical areas, some like disturbed areas (such as man-made parks, or an area after a storm), and others like grassy pastures.

Characteristics

This section of the mushroom entries contains information on the mushroom's size, color(s), shape, texture, smell, and anything else that is relevant to experiencing the mushroom.

Primary Use

The section that lists primary uses of a mushroom breaks down what (if anything) these mysterious organisms are used for by humans. There are a few major categories in which a mushroom can be placed:

- **Edible mushrooms** are used primarily for cooking. Mushrooms vary in how edible or delicious they are. Some mushrooms are poisonous or deadly, so always positively identify a mushroom (with an expert's guidance) before consuming.

- **Functional mushrooms** are used in health and wellness practices.

- **Psychoactive mushrooms** are used for hallucinogenic purposes.

- **Mycoremediation mushrooms** are useful in removing pollutants from the environment, such as in wastewater.

When a mushroom's Primary Use is labeled as "Not Applicable" (N.A.), it means either that any use for this mushroom is unknown or that the mushroom is poisonous. Those specifics are indicated in parentheses.

Moving Toward Specific Mushrooms

 Now that you have a background in the history and anatomy of mushrooms, it's time to explore seventy-five of the most intriguing mushroom species. Welcome to the mysterious and fascinating world of these unique organisms.

PART TWO

THE MUSHROOMS

27

ABORTED ENTOLOMA

Aborted Entoloma

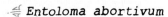*Entoloma abortivum*

AT A GLANCE

GEOGRAPHIC LOCATION + All continents except Antarctica.

GROWING LOCATION + *Armillaria* species parasitizing an *Entoloma* species, or vice versa.

CHARACTERISTICS + A white mutated globule mass growing from the forest floor. This mutated mass is actually the mushroom being parasitized.

PRIMARY USE + Edible.

The Aborted Entoloma, also called the Shrimp of the Woods, is a delicious white mushroom that grows on almost every continent. They are also called *Totlcoxcatl* (meaning "turkey wattle") in Mexico, a nickname describing the mushroom's wrinkly texture and globular irregular shape. It has no obvious look-alikes. The common name of this species has an interesting background. For a long time, scientists thought that an *Entoloma* mushroom species was the victim of a parasitic *Armillaria* mushroom species and that the parasitic relationship between the two species resulted in this mutated mushroom's growth into an odd globule structure. However, a 2001 study showed some aborted (meaning "parasitized") *Armillaria tabescens* fruiting bodies on petri dishes. Their findings showed that the *Entoloma* was parasitizing the *Armillaria*. This contradicted the previous understanding of this mushroom species. New research has only made this more confusing. Mushroom foragers around the world have found instances in the wild that make it seem like this parasitic process can swing both ways. These foragers have seen half-parasitized *Armillaria* turning into the recognizable white aborted form, but also the other way around where they find half-parasitized *Entoloma* mushrooms turning into the white aborted form. The scientific community has not reached a consensus on the order in which these changes happen, or why.

This mushroom seems to have alternating preferences for habitat. Foragers have found *Entoloma abortivum* growing around the base of trees like *Armillaria* does. However, they also grow where the *Entoloma* species more commonly grows—underneath trees, growing in the soil.

One of the most popular facets of this mushroom is its use in cooking. Because this mushroom tastes like shrimp, it can truly shine in a pasta dish. Once you get over how unusual they look and do some proper cleaning—they like to trap lots of dirt and pine needles—these mushrooms can transform any meal. One popular way to prepare them is to roast them with Old Bay seasoning and lemon. You can also panfry them until brown and add them to your favorite red sauce pasta dish. Get creative and use the Shrimp of the Woods in any dish in which you'd use shrimp.

An Abrupt Name Change

British mycologist Miles Joseph Berkeley and American herbologist Moses Ashley Curtis gave this mushroom its original Latin name (*Clitopilus abortivus*) in 1859. Around a hundred years later Dutch mycologist Marinus Anton Donk changed the name to its current moniker, *Entoloma abortivum*.

Agarikon

Laricifomes officinalis

Agarikon, also called Eburiko, Quinine Conk, or Bread of the Ghosts, is an impressively large conk (or shelf-like) mushroom that grows in layers up to 2 feet high and weighs more than 20 pounds. These mushrooms last on trees for years, growing new layers every year. Agarikon is classified as a threatened species by the International Union for Conservation of Nature, so if you chance upon it, do not harvest this mushroom. One reason for Agarikon's vulnerability is that this species mainly grows in old-growth forests, which are being wiped out by fires, logging, road construction, mining, climate change, diseases, and parasites.

Among the many historical records of Agarikon usage, the oldest evidence of human use is from a five-thousand-year-old mummy naturally preserved in the ice of the Alps. Named Ötzi after the region where he was found, this ancient man had bits of Agarikon in his stomach; scientists theorize he was using it as a food source or as a treatment for an intestinal parasite.

More than two thousand years ago, Greek physician Hippocrates referred to Agarikon as a "cure all." Widely considered to be the father of modern medicine, and authoring over seventy books in his lifetime, Hippocrates knew what he was talking about when it came to medicinal cures.

Ancient Greek pharmacist Pedanius Dioscorides also wrote about *Laricifomes officinalis* around two thousand years ago as a cure for tuberculosis. Scientists are continuing to evaluate the effectiveness of this mushroom as a natural remedy for the disease.

Though Agarikon is endangered, it is still being used for select functional purposes. Agarikon is used to support the immune system and is available in tinctures, capsules, and powders. Research into how potent this mushroom can be against diseases such as cowpox, herpes, and other viral strains is ongoing.

Outside of this mushroom's functional usage, Agarikon has many vital cultural implications. Indigenous peoples of the Pacific northwest coast of North America, including the Tlingit, Haida, and Tsimshian, referred to this mushroom as "Tree Biscuits." Shamans applied a powdered form of the mushrooms in spiritual practices or carved masks out of them for rituals and grave markings. Sometimes the fungus was woven into textiles to create a leather-type substance.

The Mushroom of Longevity

Agarikon has quite the impressive life span that may outlive an average human life span, finally rotting at 70–100 years of age.

AGARIKON

BAMBOO MUSHROOM

Bamboo Mushroom

Phallus indusiatus

AT A GLANCE

GEOGRAPHIC LOCATION +
Tropical areas.

GROWING LOCATION +
Woodlands and gardens, in woody material and rich soil.

CHARACTERISTICS + Starts as an egg-like structure, then grows a phallic white mushroom with a brown tip. It then drops down a geometric white net or veil from the top like a skirt.

PRIMARY USE + Edible and functional.

The Bamboo Mushroom, also called the Long Net Stinkhorn, or Veiled Lady, is a classic stinkhorn, or a stinkhorn that pops up in tropical areas in Asia, Africa, the Americas, and Australia. In Japan they call this mushroom *Kinugasatake*, which refers to a wide-brimmed hat with a silk veil covering the face of the *kinugasa*, or wearer. *Phallus indusiatus* was first recorded in 1798 by Étienne Pierre Ventenat, a French botanist.

A truly unique species, this mushroom is known for its veiled geometric net and foul smell. It starts its life as an egg-like structure, also called a peridium (or the mushroom's outer skin), and overnight the mushroom bursts out of the peridium. It immediately drops down a white geometric veil, called an indusium, to the ground.

The whole life of the mushroom lasts only a few days, with the fruit of this fungus disintegrating back into the soil quickly, so it's a rare sight. If you're lucky enough to find one in its egg-like form, you can put it in a Mason jar and watch it grow hour by hour. Then, watch it disappear just as quickly as it came.

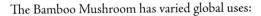

The Bamboo Mushroom has varied global uses:

+ In Mexico, it's used in divinatory ceremonies.

+ In Papua New Guinea, it's considered a sacred mushroom.

+ In Nigeria, *Phallus indusiatus* is referenced in folklore of the Yoruba, Urhobo, Ibibio, and Igbo people. Yoruba hunters also use these mushrooms as charms to make them less visible. Urhobo and Ibibio people use it in harmful charms.

+ In Asia, the Bamboo Mushroom is considered a delicacy and a commonplace functional mushroom for health and wellness. Empress Dowager Cixi of the Qing Dynasty loved this fungus so much that she used to have foragers collect the mushrooms in Yunnan Province so she could eat them in her meals.

This stinkhorn is sold dried; it can be rehydrated in simmering water to be used in soups, stir-fries, or even stuffed in poultry and cooked. This species is common in the wild, but because of its popularity, the Bamboo Mushroom has been commercially cultivated in China since 1979. China grows thousands of metric tons of these mushrooms a year. It can be grown on common agricultural wastes such as bamboo sawdust, bamboo leaves and stems, soybean pods or stems, and corn stems. In addition to its popularity as a dish, it's considered a functional mushroom, used for medicinal purposes. Research suggests that compounds in the mushroom can support the immune system, gut health, brain health, and the body's natural inflammatory response.

How the Bamboo Mushroom Spreads

When it emerges from the peridium, this mushroom is covered in a slime which has spores and foul-smelling compounds. The smell attracts insects, which unknowingly carry and deposit the mushroom's spores that get attached to them. This continues the life cycle of the mushroom.

Basket Stinkhorn

Clathrus ruber

AT A GLANCE

GEOGRAPHIC LOCATION ✦ Originally Europe, but now on all continents except Antarctica.

GROWING LOCATION ✦ Primarily woody debris, lawns, gardens, or cultivated soil.

CHARACTERISTICS ✦ Starts as an egg-like structure, then grows into a smelly, slimy reddish-orange cage shape.

PRIMARY USE ✦ Possibly edible at the egg-like stage but otherwise poisonous.

The Basket Stinkhorn, also called the Red Cage or Latticed Stinkhorn, is a memorable mushroom. This unique, extraterrestrial-looking mushroom grows in gardens, leaf litter, grassy areas, and wood chip mulches. Some people find it unsettling, as it grows quickly right in their backyard. It also emits a very strong odor, making it even less pleasant to have in one's garden. Many people will try to get rid of this mushroom as soon as it makes an appearance in their garden.

Like other stinkhorns, the Basket Stinkhorn is covered in a gelatinous slime. This slime is high in manganese, which triggers enzymes to break down keto acids and amino acids. When these enzymes break down, they create some particularly foul-smelling compounds, responsible for this stinkhorn's infamous odor. The smell, described as similar to rotting meat, attracts flies, which land on the gelatinous slime and pick up the mushroom's spores. The spores are then brought to new locations and deposited as the flies move around.

In 1560, the Swiss naturalist Conrad Gesner was the first to illustrate this mushroom, but he mistook it for a marine organism. In

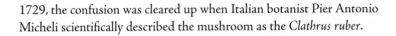

1729, the confusion was cleared up when Italian botanist Pier Antonio Micheli scientifically described the mushroom as the *Clathrus ruber*.

A Carrot's Relative?

The bright reddish-orange color of the Basket Stinkhorn comes from carotenes, primarily lycopene and beta-carotene. These carotenes are also found in tomatoes and carrots. Carotenes are useful in protecting against the sun's UV rays.

The Basket Stinkhorn has a culturally rich past. Southern European folktales warn readers of the dangers of this mushroom. As a result, many bury it when they see it. In Southern France, some considered this mushroom a cause of cancer and called it *Coeur de Sorcière*, meaning "sorcerer's heart." Similarly, in former Yugoslavia they would call it *Veštičije Srce* or *Vještičino Srce*, meaning "witch's heart." It is reported as a cause of many other ailments, such as skin rashes or convulsions.

However, in Asia and parts of Europe the mushroom "egg," or stage prior to the reddish-orange fruit, is a delicacy. The name for this dish is "deviled eggs." It's best to leave prepping these mushrooms to the experts, as there are reports of people getting sick from eating the eggs too. Proceed with extreme caution.

Mushrooms: A Gourmet Delicacy?

There are as many as fourteen thousand mushroom species in the world. Of these mushrooms:

- 50 percent are inedible.
- 25 percent are edible, but are not very appetizing.
- 20 percent will upset your stomach upon digestion.
- ~1 percent will severely harm you or kill you.

This leaves only 4 percent as delicious and desirable in the kitchen.

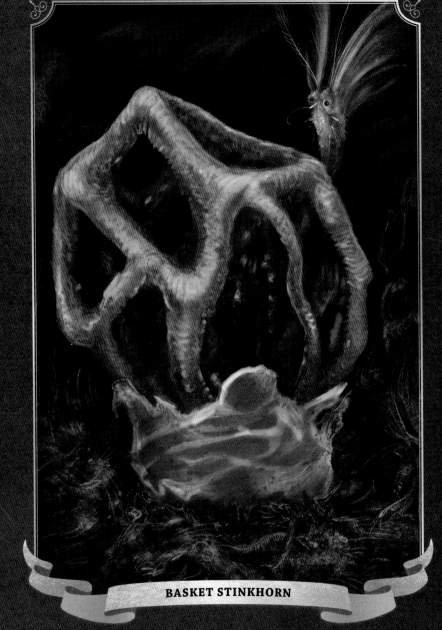

BASKET STINKHORN

BEEFSTEAK FUNGUS

Beefsteak Fungus

Fistulina hepatica

Beefsteak Fungus, also called Ox Tongue, resembles raw flesh. In the wild, Beefsteak Fungus pokes out of the side of a hardwood tree, most likely oak. It is most commonly at the base of the tree, within the first couple feet from the roots. The Beefsteak Fungus looks and feels almost like raw liver. Plus, if you squeeze this mushroom, it will exude a blood-like juice. The underside of the mushroom has a unique way to dispense spores too. When viewed under the microscope, you will see discrete tubes. These tubes look like a bunch of tiny straws where the spores are shot out of. Unlike other polypores, the porous surface of this mushroom is unique in that the pores (the small tubes to dispense spores) are not connected to each other. This mushroom's genus name, *Fistulina*, translates to "small tubes/fistules." The pore surface starts white with a light pink hue, but slowly turns reddish brown with age and bruises brown when injured. When cut open, this mushroom has an almost marbled texture, with the color mixing from red to white. This feature makes it almost identical to a cut of meat.

Unlike many mushrooms, the Beefsteak Mushroom is commonly eaten raw. The taste of this mushroom is a bit sour. This tartness is thought to be a defense against insects eating it, which is a plus for foragers: insect-free from forest to dinner table. This tangy, citrusy, sour flavor is due to its array of naturally occurring acids, including ellagic acid, malic acid, oxalic acid, aconitic acid, citric acid, ascorbic acid, and fumaric acid. A lot of chefs swear that this mushroom tastes better when raw. Creative uses include such vegetarian-friendly fare as "beef" carpaccio, "sushi" or "sashimi," plant-based "beef" tartare, kitfo (an Ethiopian dish served with spices and a teff-based flatbread called injera), and khao soi (a salad from Southeast Asia). Use discretion when eating this mushroom. In large amounts, consumption may lead to indigestion for people with weaker digestive systems.

Bird's Nest Fungi

Nidulariaceae family

(*Crucibulum* spp., *Cyathus* spp., *Mycocalia* spp.,
Nidula spp., *Nidularia* spp.)

AT A GLANCE

GEOGRAPHIC LOCATION ✦ Global.

GROWING LOCATION ✦ Decaying wood or plant matter.

CHARACTERISTICS ✦ Usually growing in clusters, this mushroom looks like a tiny bird's nest with eggs inside.

PRIMARY USE ✦ Mycoremediation, functional, or as a component of an insecticide.

The Nidulariaceae family, better known as Bird's Nest Fungi, includes the following mushroom genera: *Crucibulum*, *Cyathus*, *Mycocalia*, *Nidula*, and *Nidularia*. Since they are saprophytic mushrooms, they secrete digestive enzymes to degrade wood and plant materials for food and are commonly found in gardens or wood chip beds. They are distinctive in appearance: Each species looks like a miniature bird's nest, complete with "eggs" inside. The Bird's Nest Fungi are incredibly small; most are less than half an inch tall.

Bird's Nest Fungi were first mentioned in 1601 by Flemish botanist Carolus Clusius, and this family of fungi continues to draw scientific attention. Each species has its own peculiarities, with differences in color or use:

✦ *Crucibulum* species are tan to brownish-orange cup-shaped mushrooms. Although there haven't been any poisonous compounds found in this genus, they are not considered edible. One species, *Crucibulum laeve*, has been found to produce salfredins in liquid culture. Salfredins may inhibit an enzyme called aldose reductase (which forms cataracts in people with diabetes mellitus). What this means is that although more research is needed, salfredins could potentially benefit people with cataracts, or lead to preventive treatments.

✦ *Cyathus* species are brown to grayish brown in color; shaped like vases, trumpets, or urns; and covered in small hairs. Although not considered edible, *Cyathus limbatus* are used by Indigenous peoples in Colombia, and *Cyathus microsporus* by Indigenous peoples in Guadeloupe, for supporting sexual health. A few different types of enzymes from *Cyathus bulleri*, *C. pallidus*, *C. striatus*, *C. stercoreus*, *C. olla*, and others are used in the pulp and paper industry and the agricultural industry. These enzymes are also useful in the mycoremediation industry to break down toxic chemicals like RDX that are used in munitions. Several *Cyathus* species, including *Cyathus striatus*, are used in functional capacities to support the immune system and antioxidants.

✦ *Mycocalia* species are small barrel- to lens-shaped mushrooms with no research on edibility or uses.

✦ *Nidula* species can be white, gray, brown, yellow, or orange in color. They are cup-shaped or urn-shaped mushrooms. A compound from *Nidula niveo-tomentosa* is used in the insecticide "Cue-lure" to kill melon flies. This compound is related to the sex pheromone from female melon flies and attracts the males to their inevitable death.

✦ *Nidularia* species are irregularly shaped and look a bit different from other Bird's Nest Fungi. Their fruiting structures look almost like broken chicken eggs. Although all Bird's Nest Fungi rely on raindrops to land in the cups they produce to launch their spores, *Nidularia* species launch them sideways from their fruiting structure at speeds of up to 3 meters per second.

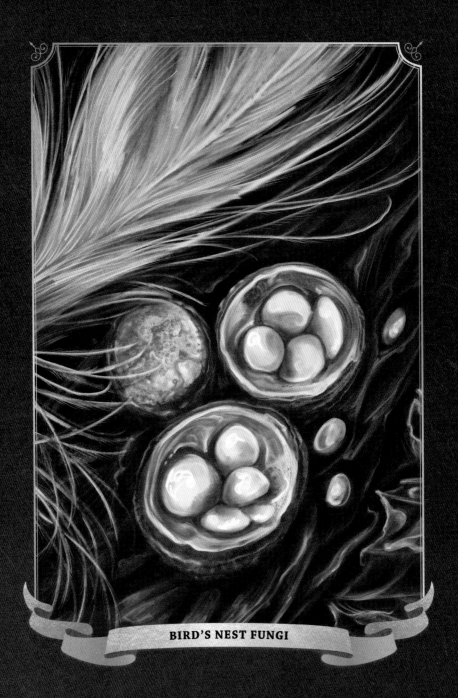

BIRD'S NEST FUNGI

BLEWIT

Blewit

Clitocybe nuda

lewits are also called Wood Blewits, or the Purple Nudist Mushroom (in Australia), and get their Latin name from their shape. *Clitocybe* means "sloping head" and *nuda* means "naked."

In the wild, these mushrooms are important to their local ecosystems because they help decompose leaf litter, mulch, and other organic matter, recycling nutrients back into the soil to feed the surrounding plants and fungi. Additionally, in Australia, male satin bowerbirds have been known to collect Blewits to decorate their structures, called bowers, to attract mates.

Some home gardeners cultivate Blewits because they create a beautiful cover crop under other plants and are a delicious edible. This mushroom can also be used to make a grass-green dye to color paper or fabrics; just chop it up and boil it in water in an iron cooking pot. For either use, gardeners typically purchase a sawdust bag of mycelium spawn (mushroom compost and mycelium) from an online vendor and mix it into a layer of straw, aged manure, vegetable compost, composted leaf mulch, bark, or wood chips either in the spring or fall. It takes about a year for the garden bed to fruit, but in North America, this usually happens in the late fall. Blewits can look like a

Cortinarius species, which can be inedible, so make sure the spore print is white and not rusty orange or brown before you consume them.

What's in a Name?

This mushroom was first described in 1790 as *Agaricus nudus*, then given the name *Agaricus bulbosus* in 1791. Its name was again changed (twice) in 1871, first to *Tricholoma nudus* and then *Lepista nuda*. The current name, *Clitocybe nuda*, was settled on in 1969. Some scientists still refer to this mushroom as *Lepista nuda*.

Whether wild or cultivated, Blewits taste great sautéed or added to creamy potato soups and are said to have a slight anise flavor. Although they are considered a choice edible by many, a small percentage of people feel slight gastrointestinal discomfort after eating these; so, when trying this mushroom for the first time, it's best to start with a small serving to see how it sits with you.

Blewits in Australia

Although Blewits are most common in Europe and North America, they have recently been introduced to Australia where they are grown with different substrates (the materials that the mushrooms grow in). These new substrates are changing both the appearance and the growth pattern of the Blewit. Pretty soon this Australian variety might be given yet another new Latin name if it diverges enough.

Blue Pinkgill

Entoloma hochstetteri

Blue Pinkgill, also known as the Sky-Blue Mushroom, is an indigo-blue mushroom with a slight tint of green. The caps of this small mushroom grow to a maximum diameter of about 1.5 inches. Although this mushroom is typically all blue, the gills can sometimes have a slight reddish-pink color, making it seem as though the spores are pink. If you take off the cap, place it on paper or tinfoil, cover it with a cup and let it sit overnight, when you lift the cap the next day you will see a beautiful pink spore print pattern.

The Blue Pinkgill was originally called *Cortinarius hochstetteri* in 1866 by Austrian mycologist Erwin Reichardt. This mushroom was given its current scientific name in 1962 by New Zealand mycologist Greta Stevenson. It's named after the German-Austrian naturalist Ferdinand von Hochstetter, who originally described and drew the mushroom. The Pinkgill is called *Werewere-Kokako* by the Māori after the kōkako bird, as this bird has blue wattles. The story is that the kōkako bird got its blue wattles by rubbing its cheeks on the Blue Pinkgill Mushroom; some of the blue pigment rubbed off and transferred onto the bird permanently. The Māori have traditionally used this mushroom for tattoo ink, food, and medicine.

The mushroom gets its bright blue color from three azulene pigments in the mushroom. These pigments are now being studied for potential uses as food dyes; however, it is currently unknown whether this mushroom is toxic, or if it has psychoactive properties.

New Zealand's Favorite Mushroom

One of the most popular mushrooms in New Zealand, the Blue Pinkgill made its appearance on a fungal-themed postage stamp in 2002 and is on the New Zealand fifty-dollar note. In 2018 it was ranked as the first pick for New Zealand's national fungus.

Mushrooms Are a Small Part of the Fungi World

There are about six times as many fungus species as plant species in the world. Estimates vary about how many fungus species exist—anywhere from about two million to six million. But only about a hundred and forty thousand species of fungi have been categorized by scientists. And of these, only around fourteen thousand produce mushrooms.

BLUE PINKGILL

CAESAR'S MUSHROOM

Caesar's Mushroom

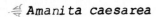

Amanita caesarea

AT A GLANCE

GEOGRAPHIC LOCATION ✦ Europe, North Africa, India, Iran, China, and Mexico.

GROWING LOCATION ✦ Primarily oak forests.

CHARACTERISTICS ✦ Classic *Amanita* cap-and-stem mushroom growing from a white egg-shaped structure. The stem is a blend of yellow, orange, and white and the cap has a reddish-orange color.

PRIMARY USE ✦ Edible with caution.

This mushroom was originally named *Agaricus caesareus* in 1772 by Italian naturalist Giovanni Antonio Scopoli. However, in 1801 German mycologist Christiaan Hendrik Persoon gave it its current name, *Amanita caesarea*. Caesar's Mushroom gets its common name from Roman emperor Tiberius Claudius Caesar Augustus Germanicus, who loved eating this mushroom so much that it was his eventual downfall. He mistook a deadly look-alike—the Death Cap—for this mushroom, and it killed him. Some have claimed that it was an assassination rather than an accidental misidentification, with various rumors pointing to his taster, Halotus; his doctor, Xenophon; the infamous poisoner Locusta; or even his wife, Agrippina. We will never know what really happened, but what we do know is that Emperor Claudius sure did love eating plates full of this delicious mushroom.

Caesar's Mushroom varies greatly in size: The cap can be anywhere from 2.4–5.9 inches in diameter while the stem is 3.1–5.9 inches in diameter. Growing primarily in oak forests, but sometimes amid conifers, in Europe, North Africa, India, Iran, China, and

Mexico, this mushroom is a mycorrhizal species, meaning it has a symbiotic relationship with the tree roots. In Italy, this mushroom, which grows out of an egg-like puffball structure, is called *Ovolo Buono*, which translates to "good egg." The taste of the mushroom is light, with a mild nutty flavor (like a hazelnut or chestnut). In Italy and Sicily, they cook this mushroom by slicing it and rolling it in salt, lemon juice, olive oil, and pepper.

Craving a Mexican Recipe?

In Mexico, Caesar's Mushroom is routinely cooked and roasted with an herb called epazote. This herb has a pungency similar to fennel, oregano, anise, or even tarragon, and is sometimes called "Mexican tea."

When harvesting puffballs, which can look like the egg-like stage of a mushroom, it's always advisable to cut the puffball in half to see if there is really a mushroom inside. This puffball could potentially be mistaken for deadly look-alikes like the Death Cap or Destroying Angels, which also grow from "eggs." Caesar's Mushroom has a few other mushroom doubles. In North America, there is a related species called *Amanita jacksonii*, which is also edible and looks very similar. It may also be mistaken for the psychoactive Fly Agaric, especially when the Fly Agaric loses its defining attribute: its white scales on its reddish-orange cap. The Fly Agaric Mushroom is unsafe to eat, as it can make you sick when ingested.

Mushroom and Human DNA

Mushrooms are often mistaken as plants, but that's far from the truth. Mushrooms actually have more in common with humans than they do with plants. We even share about 50 percent of our DNA with mushrooms.

Caterpillar Fungus

Ophiocordyceps sinensis

AT A GLANCE

GEOGRAPHIC LOCATION ✦ China, Nepal, India, Bhutan on the Tibetan Plateau, and the Himalayas.

GROWING LOCATION ✦ Parasitizes caterpillars of the ghost moth at about 11,000 feet above sea level.

CHARACTERISTICS ✦ Brown stick-like mushroom grows out of the head of a ghost moth caterpillar.

PRIMARY USE ✦ Functional.

Formally known as *Cordyceps sinensis*, this mushroom is one of the rarest and most expensive mushrooms in the world. The common English name is Caterpillar Fungus, but it is known in China as *Dōng Chóng Xià Cǎo*, which translates to "winter worm, summer grass." This interesting mushroom is also commonly known in Tibet as *Yartsa Gunbu*, also translating to "winter worm, summer grass."

Growing at high elevations (about 11,000 feet above sea level) and barely noticeable in the grass, this can be an incredibly elusive mushroom. Combining the fact that it's incredibly revered in traditional Chinese herbalism with the fact that it's incredibly difficult to find, means these mushrooms come with a hefty price tag for quality specimens. There are entire families and communities that base their livelihoods on harvesting this mushroom. Harvesting rights are highly contentious, leading to much bloodshed. Disputes over this mushroom even had an impact in the Nepalese Civil War: Nepalese Maoists fought government forces over control of trade during the peak harvest season. Some sellers have been known to put lead weights into the mushrooms so they would weigh more and get a higher price. Since the harvesting is so cumbersome and expensive, some people have opted to grow the mycelium in bioreactors as a source of compounds. Others grow a similar species called *Cordyceps militaris*, which has a lot of the same compounds. The author, Alex Dorr, built the largest and first USDA-certified organic *Cordyceps militaris* mushroom farm in the Americas in 2019. To date, only a couple people have figured out how to grow the Caterpillar Fungus indoors, but not at a large scale. The life cycle of this mushroom is difficult to manage. So, people must rely on foraging for this rare mushroom in the wild.

An entomopathogenic fungus, meaning that it parasitizes insects, the Caterpillar Fungus starts its life cycle as an endophytic fungus living in the roots of native plants. Then, when the ghost moth caterpillars eat the roots of the plants, the caterpillars become infected. The fungus takes over the insect's body and positions itself just under the soil so the mushroom can grow. The mushroom then spreads its spores to continue the life cycle. It's best known for a compound that research has shown to be beneficial in supporting energy levels. It's used traditionally in teas, but can also be made into tinctures, powders, or capsules. It can also be cooked into a broth or stuffed into chicken, turkey, or duck before cooking.

Harvesting Caterpillar Fungus

For those who rely on the Caterpillar Fungus for income, it can be a tough job. The stalks of the mushrooms are similar in appearance to native grasses. Also, digging the caterpillars out of the ground requires precision: If the stalk is disconnected from the caterpillar, it becomes much less valuable. All this, coupled with the dense Tibetan soil, makes for a difficult profession.

CATERPILLAR FUNGUS

CAT'S TONGUE

Cat's Tongue

Pseudohydnum gelatinosum

AT A GLANCE

GEOGRAPHIC LOCATION ✦ Potentially global.

GROWING LOCATION ✦ Dead conifer trunks, logs, and stumps.

CHARACTERISTICS ✦ Small, translucent, gelatinous, and tongue shaped. This mushroom has tiny teeth on the underside of the cap that feel like the tongue of a cat.

PRIMARY USE ✦ Edible and functional.

Cat's Tongue, also known as Toothed Jelly Fungus, False Hedgehog Mushroom, or White Jelly Mushroom, has arguably the most interesting touch sensation in the fungal kingdom. These mushrooms are known for their unique textures, making them a fun tactile experience for new mushroom hunters and kids. Touching the underside of this mushroom is almost identical to a cat licking you. Like a lot of jelly fungi, Cat's Tongue has a unique ability to shrivel up in dry weather and reanimate when there is enough rain or moisture in the air suitable for spore dispersal.

Cat's Tongue was originally named *Hydnum gelatinosum* in 1772 by Italian naturalist Giovanni Antonio Scopoli. It was given its current Latin name, *Pseudohydnum gelatinosum*, by Finnish mycologist Petter Adolf Karsten in 1868. *Pseudohydnum* means "similar to the genus *Hydnum*" and *gelatinosum* refers to the gelatinous jelly-like texture of this mushroom. Cat's Tongue is the only toothed jelly mushroom, meaning it has tiny tooth-like structures on the underside of the cap where spores come out.

Although for the last couple hundred years this mushroom was thought to be a globally occurring species, new DNA analysis has found that true *Pseudohydnum gelatinosum* only exists in Europe and northern Asia. The other species that were once included under the Latin name *Pseudohydnum gelatinosum* in Australia, New Zealand, North America, Central America, and South America are still being described and separated into their new Latin names, but more research needs to be done. For example, we now know there are three distinct species in North America, but they have not yet been named, which is why they are all still called *Pseudohydnum gelatinosum*.

Although most people don't bother with this species since it's so small and won't yield much food, it can be turned into some very interesting creations. Apparently, it's amazing with honey and cream, made into a marinade to be used in salads, or even used to flavor ice cream. Studies show that *Pseudohydnum gelatinosum* is effective for supporting the immune system, since it's packed with polysaccharides and lectins. This immune system support is important for people with busy lives, seasonal transitions, lots of travel, and more, to keep your body in tip-top shape.

Cat's Tongue Candies

Some curious and culinarily adept mushroom lovers have discovered that you can make vegan gummies out of these fun mushrooms. Although the name doesn't sound vegan, Cat's Tongue gummies, prepared the right way with only a few necessary supplies, will satisfy your pals who stick to a plant-based diet.

Cauliflower Mushroom

Sparassis crispa

The Cauliflower Mushroom, or *Hanabiratake* as it's known in Japan, is a famous gourmet and functional mushroom that grows around the world. It is commercially grown and sold in Korea, Japan, China, the United States, and Australia. The name *Sparassis* comes from the Greek word *sparassein* ("to tear") while *crispa* comes from a Latin word meaning "curly." To some this mushroom looks like a head of cauliflower, hence its common name. It also looks like a sea sponge, and can grow massive in size, in clumps weighing up to 66 pounds. If left on a dark piece of paper, it will produce a white spore print.

Preparing this mushroom for eating can be a lot of work. Its interesting woven shape can trap lots of sticks, dirt, leaves, and bugs, so clean this mushroom carefully and thoroughly. The initial scent of the mushroom is of ammonia and latex, but that scent quickly gets cooked out. Also, to most enjoy the Cauliflower Mushroom, you may need to cook it for longer than anticipated. Otherwise, it will have a chewy, rubbery consistency. One delectable recipe for this mushroom is to cook it for one hour in chicken, venison, or duck broth, and then deep-fry the leaflets with a tempura batter for a crispy, delicious experience.

Other chefs call for beef pot roasts, pickles, or sautéing them in a pan for a long time to cook down the rubbery consistency.

As a functional mushroom, *Sparassis crispa* is packed with a load of beneficial compounds. This mushroom continues to be extensively researched and is used to:

+ support the immune system
+ help regulate a healthy and natural inflammatory response after a workout
+ keep blood sugar levels in normal ranges
+ promote skin health
+ support your body in seasonal transitions

Dried Cauliflower Mushrooms have been shown to contain a large amount of a compound that supports the immune system. Since this is such a sought-after mushroom for both its functional benefits and its uses in the kitchen, it's been commercially grown all around the world. This is great for people who don't want to go out in the wild and search under the conifers, which might not exist in their region anyway. Since cultivators can grow these mushrooms in clean environments, it solves the issue of organic debris getting lodged in the mushrooms' intricate wave-like flesh.

Foraging Cleaning Tips

Unfortunately sticks and mud are not the only things that may reside in your mushrooms. If your finds are freshly foraged, there's a chance you will encounter insects or maggots. To ensure cleanliness, cut out discolored or hole-ridden mushroom flesh. You will want to do this before bringing a mushroom into your home.

CAULIFLOWER MUSHROOM

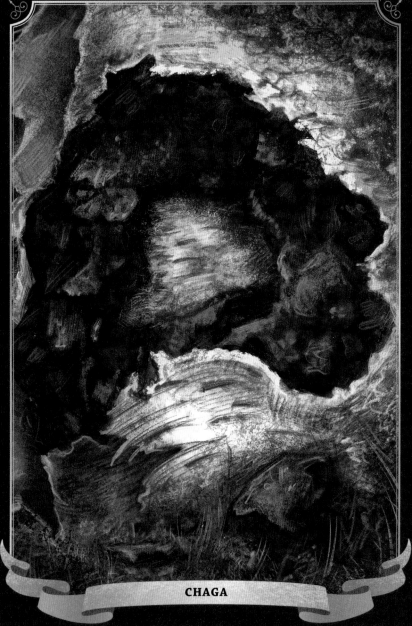

CHAGA

Chaga

Inonotus obliquus

*N*ot technically a mushroom, Chaga is considered a conk, canker, or sclerotium, which basically means that it is a hardened mass of mycelium and wood. Some call it Clinker Polypore or Gold of the Forest because of the golden color in the center of this mushroom. Chaga has other names too. In Norway they call in *Kreftjuke* (meaning "cancer fungus"), in Finland, *Tikkatee* (meaning "darts"), and in Japan, *Kabanoanatake* (meaning "Betula," referring to the genus of birch trees). Cree healers called it *Posahkan* or *Wisakecahk Omikih* (which refers to a mythical character named *Wisakecahk* who threw a scab on a birch tree which then sprouted Chaga). The Cree used it as incense, tinder to start fires, or in a pipe to keep the tobacco lit. During World War II, Finnish people drank a Chaga drink instead of coffee.

Caution should be taken to not overharvest Chaga, since at the stage at which it is harvested, it has not yet spread its spores. Overharvesting can threaten the prevalence of this species. When people do harvest it, usually it's done in the winter with snowshoes, a backpack, and a pick or axe to get the hard, charcoal-like conk off the tree. The Gold of the Forest can sometimes grow as high as 30 feet up a tree, which can make harvesting a challenge.

How Does This Fungus Get Its Color?

Just as humans have darker skin colors when their skin produces more melanin, the same is true in other species. The outer color of this dark, almost charcoal-like fungus is a result of its high melanin content.

Dried forms of Chaga are easily accessible online, usually in tea mixes. This is due to its history. The traditional way to ingest Chaga was to make a hot tea, which makes a delicious and earthy vanilla chai flavor. Today, it's often mixed with coffee, chai lattes, and hot chocolate. Chaga is packed with compounds that are incredible for supporting the immune system, which is yet another reason for its frequent use. Besides supporting the immune system, Chaga is found in many skin products because the high amounts of melanin present in Chaga help support healthy skin. This fungus is also known for offering:

- ✦ potential antioxidant support
- ✦ support for healthy aging
- ✦ adaptogen benefits (helping the body adapt to occasional stress and fatigue)

Although Chaga tea is a delicious drink, like all functional mushrooms, it's best to use a dual extraction, meaning using both alcohol and hot water to access all the beneficial compounds. If one is using tea, the great thing about Chaga is that you can use the same chunk of Chaga to brew a pot of tea every day for months and still pull incredibly dark and rich tea from it. A little bit goes a long way.

Chanterelles

Cantharellus **spp.**

There are about a hundred different species of Chanterelles in the *Cantharellus* genus. Many of these mushrooms are highly sought after for their culinary delights. Only available in the wild in the summer and fall, they are coveted by foragers. The name Chanterelles comes from the Greek word *kantharos*, meaning "tankard cup," which refers to the mushroom's shape.

The cooking potential with Chanterelles is practically endless: They can be paired with various meats and fish, and used on pizzas, in pasta, in crepes, in stir-fries, in egg dishes, in sauces, or even pickled in brine. Keeping it simple by just panfrying them with a few spices is also a delicious option; they taste of black pepper or slightly peachy. Another amazing way to consume these mushrooms is by caramelizing them in maple syrup and using them as a topping on ice cream. Chanterelles are said to have a faint aroma of apricots, which is extremely pleasant when you have a basketful.

Look-alikes include the False Chanterelle (*Hygrophoropsis aurantiaca*), which although also edible, will disappoint you if you are thinking it's a true Chanterelle. A dangerous look-alike are the Jack O'Lantern Mushrooms (*Omphalotus illudens*, *O. olearius*, *O. olivascens*, and *O. subilludens*), which although not deadly, will make you very sick. So, as always, exercise caution while foraging.

Here are some interesting Chanterelle species:

+ *Cantharellus cibarius* is the type species for this genus, meaning it's the most popular species in the genus. This species is where the genus's name comes from. It's called a Golden Chanterelle or Girolle. It grows in Europe in deciduous and coniferous forests.

+ *Cantharellus incrassatus* was only found once in the Pasoh Forest Reserve in Malaysia, and never found again. There could be other *Cantharellus* species in this park reserve being protected inadvertently that are yet to be discovered.

+ *Cantharellus minor* is a tiny Chanterelle mushroom that can even have symbiotic mycorrhizal relationships with moss. They were first reported only in North America but recently have been reported in Kerala, India. How they showed up there, or whether they grow elsewhere, is still a mystery.

+ *Cantharellus persicinus*, also called the Peach Chanterelle or the Pink Chanterelle, is exactly how it sounds—peachy-pink colored. It grows in the Appalachian region of North America and is a delicious edible mushroom.

+ *Cantharellus subalbidus*, also called the White Chanterelle, is white instead of the typical orange-yellow color. Surprisingly enough, it bruises orange.

+ *Cantharellus lilacinus* is a beautiful purple- to lilac-colored Chanterelle with large, wide gills found in Australia; it was first described in 1919.

+ *Cantharellus texensis* is a fiery reddish-orange Chanterelle found primarily in Texas, but it has popped up in Louisiana, Mississippi, Alabama, and Florida.

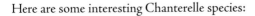

🍄 How to Spot a Fake Chanterelle

While not obvious, there is one telltale sign that you are looking at a fake Chanterelle rather than a genuine, delicious one. All Chanterelles have light-colored folds underneath their tops that are not true mushroom gills. Fake Chanterelles have individually moveable, true gills that are deep tan in color.

CHANTERELLES

CHICKEN *of the* WOODS

Chicken of the Woods

 Laetiporus sulphureus

Chicken of the Woods, Sulphur Shelf, or Sulphur Polypore is one of the most highly sought-after edible mushroom species because it can be cooked to taste exactly like chicken. Chicken of the Woods was originally described in 1780 by French mycologist Jean Baptiste François Bulliard. Its current Latin name came in 1920 from American mycologist William Murrill. With more DNA analysis, there may be a Latin name change on the horizon for any Chicken of the Woods found outside of Europe, as the species change slightly. Stay tuned, as Latin names for mushrooms change all the time.

Chicken of the Woods grows in Europe and North America. It causes cubical rot of a variety of trees. Cubical rot is when the fungus breaks down the wood and it cracks into cubical pieces. This mushroom is one of the easiest identifiable species for beginners because of its striking features. The only potentially dangerous look-alike would be the Jack O'Lantern Mushrooms, which are poisonous, orange-clustered mushrooms that grow at the base of trees. However, the look-alike Jack O'Lanterns are cap-and-stem mushrooms with gills and Chicken of the Woods are shelf mushrooms with pores.

Record-Setting Mushroom

Chicken of the Woods set a Guinness World Record when the world's heaviest edible fungi specimen was found on October 15, 1990. It weighed 45.35 kilograms (100 pounds) and was found in the New Forest in Hampshire, United Kingdom, by Giovanni Paba.

Those with sensitive stomachs should be wary of eating this mushroom. Around 10 percent of the population feels gastrointestinal issues when ingesting Chicken of the Woods. Research has shown that this stomach discomfort comes from eating two subspecies of the Chicken of the Woods: *Laetiporus huroniensis* on hemlock and *Laetiporus gilbertsonii*. If you have never eaten Chicken of the Woods before, it's recommended to start with a very small quantity to see how it sits with you before eating a large amount. There is a similar phenomenon with Morels and a few other highly sought-after gourmet mushrooms, where a small percentage of the population has a slight adverse gastrointestinal reaction.

Although true Chicken of the Woods is *Laetiporus sulphureus*, there are other edible *Laetiporus* species that look extremely similar that people also call Chicken of the Woods and eat just the same; examples include:

- *L. cincinnatus*
- *L. huroniensis*
- *L. gilbertsonii*
- *L. persicinus*
- *L. conifericola*

These species extend into Asia, the Caribbean, Australia, and South America.

Although this mushroom tastes great sautéed, you can be very creative with how you cook it. There's even a vegan remake of a famous fast-food chicken chain's chicken sandwich made with breaded and fried Chicken of the Woods mushrooms replacing the chicken. People say that this recipe is mouthwatering and a "must make" if you get your hands on Chicken of the Woods and an air fryer.

Coral Fungus

Ramaria spp.

Coral Fungus is part of the genus *Ramaria*, which houses approximately two hundred species. Coral Fungi are also known as Clavarioid Fungi. T. Hjomsköld introduced the name *Ramaria* in 1790, and between 1821–1933 it was considered a section of the genus *Clavaria* until Dutch mycologist Marinus Anton Donk put it in its own genus.

Of the many species of *Ramaria*, there are a few that stand out:

✦ *Ramaria flava* is an edible species of yellow color, found in Europe, southern Chile, and Brazil. As with most mushrooms, it's best to eat this species after it's cooked. (In this case, it should be cooked for at least fifteen minutes.)

✦ *Ramaria formosa* is a poisonous mushroom found in North America, Asia, and Europe. This species of mushroom is also called Salmon Coral, Beautiful Clavaria, Handsome Calvaria, or Pink Coral Fungus. Although this mushroom is considered mildly poisonous, giving people nausea, abdominal pain, and diarrhea, some people love to eat it. Fans of this mushroom say that it is safe to eat if you remove the acrid tips. This mushroom is even thought to be a functional ingredient in traditional herbalism by the Gurjar and Bakarwal tribes in the Rajouri and Poonch districts of India. You may also find this mushroom sold in markets in Lijiang, China. Despite this anecdotal evidence, this mushroom is still considered poisonous. So, maybe leave the consumption to the experts.

✦ *Ramaria botrytis* is another edible *Ramaria* species, commonly known as the Pink-Tipped Coral Mushroom or the Cauliflower Coral, which grows in Africa, Australia, Chile, Asia, Europe, Mexico, and Guatemala. The taste of this mushroom seems to vary. It has been described as being fruity, or similar to the taste of sauerkraut, green peanuts, or pea pods. In central Italy, they put Cauliflower Coral in a stew or pickle it in oil. However, eat this mushroom with caution. Some people advise against eating it, as it may have laxative effects. Beginning foragers might confuse it with the *Ramaria formosa* species and may eat a poisonous mushroom. If you are willing to eat the mushroom regardless of any potential negative gastrointestinal effects, it's worth knowing that *Ramaria botrytis* is used as a functional mushroom able to support the immune system.

All in all, Coral Mushrooms in the *Ramaria* genus are beautiful to chance upon, and with their variety, they are an amazing group of mushrooms to learn more about.

Oceanic Mushrooms

Researchers off the coast of Japan took rock samples in the Pacific Ocean and found sixty-nine types of fungi living in hibernation underwater. One of the species was found to have been in hibernation for twenty million years. After it was harvested, this species not only germinated to grow mycelium on an agar media, but also fruited mushrooms.

CORAL FUNGUS

CORDYCEPS

Cordyceps

Cordyceps militaris

AT A GLANCE

GEOGRAPHIC LOCATION ✦ All continents except Antarctica.

GROWING LOCATION ✦ Primarily on moth cocoons or pupae, which are nuzzled in moss or leaf litter.

CHARACTERISTICS ✦ Similar in appearance to a cheese puff, this mushroom is an orange, club-shaped ascomycota (or a type of fungi that are known as sac fungi because they have tiny sacs, called perithecia, where the spores are held). These mushrooms grow out of the soil or moss on a black cocoon or pupa.

PRIMARY USE ✦ Functional and edible.

Cordyceps, also called the Scarlet Caterpillar Club Mushroom, was first named in 1753 by Carl Linneaus, the father of taxonomy, after he discovered them growing in the wild. Native to all continents except Antarctica, this pervasive mushroom can also be cultivated artificially on a farm for grocery or health food needs. First grown in the United States by Ms. Green at Cornell University in 1894, this mushroom has become wildly popular with professional cultivators and home growers. Today, it is primarily grown in Asia at a capacity of hundreds of millions of dried kilos a year.

This mushroom is classified as an entomopathogenic fungus, meaning it attacks and feeds on insects in the wild. However, the most popular cultivation method of this mushroom is vegan, using rice as a growing medium in jars or trays. In the wild, Cordyceps are mostly found in damp areas, likely by a stream, marsh, or river bed. They like to grow in moss or leaf litter in areas with high humidity. The stromas, or stalks, grow to a maximum height of about 3 inches.

Due to their small size, Cordyceps can be extremely hard to spot when foraging for them. Normally, you will find one mushroom per moth cocoon, but sometimes you can find up to eight mushrooms popping out from one cocoon. This mushroom is ready to harvest when the tip of the stroma is covered in tiny orange bumps called perithecia, which are sacs that hold the spores. Be careful when digging out the cocoon that's buried underground, as it's used to properly identify the mushroom. If you were to break open this cocoon, the inside would be a dense, whitish mass of mycelium where the former pupa used to be. Once picked, these mushrooms can be dried and sent in for DNA analysis or sent to a mushroom farm to culture the mushroom tissue resulting in more mushrooms.

In traditional Chinese herbalism and Western herbalism, this mushroom is primarily used for supporting energy levels. Packed with great compounds, this mushroom helps support natural energy production on a cellular level. Praised by athletes worldwide, and a fan favorite for chefs wanting to make a yummy Cordyceps soup, this is a highly sought-after mushroom.

People will typically ingest this mushroom in a dual-extracted tincture, powder, or capsule. Making a tea or soup broth is a delicious option, but you only get a fraction of the beneficial compounds in this way. The most common part of the fungi used in herbalism is the actual mushroom fruiting bodies—not the bug or the mycelium. Cordyceps Mushroom's fruiting bodies have many more benefits than the other parts of the fungus.

Other Helpful Benefits of the Cordyceps

As an adaptogen, Cordyceps helps the body fight off occasional stress and fatigue and supports a person's natural immune response. Helping to support lung capacity, stamina, and natural drive, this is a favorite for whatever activities are at hand that might cause you to build up a sweat.

Darwin's Fungus

Cyttaria darwinii

Darwin's Fungus, also called the Beech Orange, was named after Charles Darwin. Darwin collected this fungus in Tierra del Fuego, South America, during the voyage of the HMS *Beagle* in 1832. This species was first scientifically described by British mycologist Rev. Miles J. Berkeley in 1842. Berkeley then named it after Charles Darwin.

Since this fungus only grows on southern beech trees, it is found only in the Southern Hemisphere (mainly in Argentina and Chile). The Beech Orange infects the tree and triggers a reaction in it, causing the tree to form a gnarled gall. The mushroom then sprouts out of these galls. Craftspeople sometimes use these gnarled galls to make bowls and other objects. After shooting their spores out of the dimples (perithecia) on the surface, the mushrooms drop from the tree, leaving the gnarled gall on the surface behind. Since the fungus doesn't cause the tree much harm, it's still not understood whether this is a parasitic (one benefits while the other is harmed), commensal (one benefits while the other is neither harmed or benefits), or mutualistic (they both benefit) relationship between the fungus and the tree.

Darwin's Fungus is known primarily as an edible mushroom. These mushrooms are a delicacy to the Indigenous people of Tierra del Fuego, Darwin wrote. The scientist saw women and children collect vast amounts of these mushrooms to eat. According to Darwin, the mushrooms, traditionally eaten raw with fish, had a sweet but mild taste.

There are a couple of closely related species including *Cyttaria harioti*, also called *Llao Llao* or *Pan de Indio*, which looks very similar to *Cyttaria darwinii* although a bit more orange in color. *C. harioti* is used by the Araucanian people of Chile in a fermented drink they call *chica del llau-llau*. Since this mushroom contains up to 15 percent fermentable sugars and is coated in the same *Saccharomyces* yeast that many alcoholic drinks are made with, the Araucanians can dry, grind, and mix these mushrooms with warm water and ferment it into an alcoholic beverage. The second closely related species is *Cyttaria gunnii*, which are eaten in abundance by Indigenous peoples in Australia and New Zealand. It is very similar in appearance to Darwin's Fungus.

Fungi Underground

Fungi make up approximately 90 percent of the total microbial biomass of forest soils on this planet. Fungi's vast underground networks connect almost all the trees and plants in forests, helping plants communicate and trade nutrients. They also store huge amounts of carbon, creating the greatest carbon sink that we know.

DARWIN'S FUNGUS

DEAD MAN'S FINGERS

Dead Man's Fingers

Xylaria polymorpha

Dead Man's Fingers is a mushroom right out of a horror movie. It looks exactly as it sounds, like a ghostly hand popping up from the soil. Sometimes they grow on hardwood logs aboveground, but when the wood is buried and the mushrooms are in the right formation, they really do look creepy. Dead Man's Fingers only grows on already injured or deadwood of hardwood trees such as beech or maple. When this mushroom is young, it has a whitish-bluish growing tip with a darker-colored body. Then, as it grows older, it becomes black and covered in small bumps (called perithecia), from which the spores are released.

As with many species of mushrooms, Dead Man's Fingers can be "species complex," meaning there are many look-alikes that, without DNA sequencing, makes it almost impossible to differentiate between them. It's estimated that less than 10 percent of the fungi in the world have been described; as more mycological research is done, more species will be given individual names.

In Ayurvedic herbalism in India, Dead Man's Fingers Mushrooms are known as *Phoot Doodh*, meaning "to gush milk." They are used to promote lactation after birth and taken twice daily before meals with cow's milk. For those that do consume the mushrooms, there appears to be no distinct taste or smell. People should be wary of consuming these mushrooms, though. Twenty-eight species of *Xylaria*, including *Xylaria polymorpha*, were tested for toxins and they all came back positive for having two different types of toxins. This is counterintuitive since many people around the world eat this mushroom and use it for functional wellness, without reporting any signs of toxicity. More research needs to be done on this mushroom species and genus before ingesting.

Though it may be partially poisonous, it does seem to have benefits. Research shows this mushroom is packed with beneficial compounds used to support the immune system. In lab settings, researchers were able to produce beneficial compounds which are widely used in the mycoremediation industry for degrading toxic wastes in the environment such as monochlorotriazine dye.

Musical Mushrooms

Francis Schwarze, a professor from Empa, the Swiss Federal Laboratories for Materials Science and Technology in St. Gallen, Switzerland, used a species related to Dead Man's Fingers, *Xylaria longipes*, when making a replica of a Stradivarius violin, which can cost upward of $20 million. He used the fungi to create microfractures and channels in the wood, creating a more vibrant sound. In a double-blind study, participants actually liked the sound of the fungal violin a lot more than the sound of the multimillion-dollar Stradivarius. Sounds like we need more fungal instruments.

Deadly Galerina

Galerina marginata

*D*eadly Galerina, also known as Funeral Bell, Deadly Skullcap, or Autumn Skullcap, is one of the most infamous poisonous mushrooms in the world. Because these species look so different, it's also one of the hardest to identify. The Funeral Bell is so difficult to identify that members of the same genus, including G. *autumnalis*, G. *oregonensis*, G. *unicolor*, G. *venenata*, and G. *pseudomycenopsis*, were all thought to be from totally different genera. This is because these mushrooms all presented different structural and ecological characteristics. Mycologists had to go one step further: DNA testing. When DNA analysis was brought in, researchers realized these different-presenting species were all under the same genus and most of these "different" mushrooms were lumped under the Latin name *Galerina marginata*.

Deadly Galerina was first called *Agaricus marginatus* in 1789 by German naturalist August Batsch; its name was later changed to *Galerina marginata* by American mycologist A.H. Smith and German mycologist Rolf Singer in 1962. This mushroom is lumped into a large group of hard-to-identify mushrooms called "little brown mushrooms" otherwise known as LBMs. For this reason, it's best to avoid any little brown cap-and-stem mushrooms growing off wood as they could very well be Deadly Galerina. Falling for the trap of the LBM has been the downfall of a few overzealous mushroom hunters. They may have thought they had found psychedelic Psilocybin species, the highly edible Honey Mushrooms, or Sheathed Woodtuft Mushrooms. The Deadly Galerina can also be confused for the edible *Flammulina velutipes* mushroom, or even misidentified as another toxic mushroom that contains the same toxins, *Conocybe filaris*.

Galerina marginata have the same toxins as the notorious Death Cap Mushroom (*Amanita phalloides*) and the Destroying Angel species (*Amanita bisporigera*, *A. ocreata*, *A. virosa*, and *A. verna*). Certain compounds found in *Galerina marginata* inhibit an enzyme that is essential for cells to function properly in your body. Once cell metabolism stops, your organs start to fail. Around 15 percent of those poisoned by *Galerina marginata* will die within ten days; other common reactions include liver, kidney, and respiratory failure and coma. In short, always positively identify your mushrooms before consuming them—and don't hunt for LBMs.

A Special Galerina

Despite having DNA that's practically indistinguishable from *Galerina marginata*, *G. pseudomycenopsis* is different in a few key ways: the species grow very far apart physically, and these mushrooms are unable to mate with each other. Therefore, *G. pseudomycenopsis* was given a separate Latin name and considered a separate species.

DEADLY GALERINA

DEATH CAP

Death Cap

Amanita phalloides

AT A GLANCE

GEOGRAPHIC LOCATION ◆ North America, Europe, North Africa, and the Middle East.

GROWING LOCATION ◆ On the ground, in mycorrhizal association with trees.

CHARACTERISTICS ◆ This cap-and-stem mushroom has a white cap with a green, tan, and yellow hue. It sprouts from an egg-like structure and has a partial veil on the stem.

PRIMARY USE ◆ N.A. (potentially fatal if ingested).

The infamous Death Cap Mushroom is the deadliest mushroom on the planet, resulting in most of the mushroom poisoning deaths in the world. It was first described by French botanist Sébastien Vaillant in 1727, but it was later given its current Latin name, *Amanita phalloides*, by German naturalist Johann Heinrich Friedrich Link in 1833. This mushroom is often grouped alongside the deadly Destroying Angel mushroom species (*Amanita bisporigera*, *A. ocreata*, *A. virosa*, and *A. verna*) but the *Amanita phalloides* is in a category of its own in terms of toxicity and has earned its Death Cap name.

The Death Cap was responsible for some of the most infamous poisonings in history: Roman Emperor Claudius, possibly Pope Clement VII, the Russian tsaritsa Natalya Naryshkina, and Roman Emperor Charles VI. What makes this mushroom doubly dangerous is that it looks like some common edible mushrooms. Roman Emperor Charles VI's favorite dish involved the Caesar's Mushroom, *Amanita caesarea*. These mushrooms are very similar in appearance.

And, to make things even worse, the Death Cap also looks like another of the most commonly consumed mushrooms in the world, the Paddy Straw Mushroom (*Volvariella volvacea*). This is especially dangerous for people who grew up eating the Paddy Straw Mushroom in Southeast Asia. If these people then move to a different country where the Death Cap Mushroom grows in the wild, they may get excited, harvest and cook with the Death Cap, and then poison their family and friends. People who *have* eaten this mushroom claim it tastes really good, and symptoms don't kick in until a few days after ingesting it, making treatment incredibly hard. Usually when people are poisoned, symptoms are almost instant, so they can be treated quickly. With the Death Cap, the delay of symptoms makes it hard for the ER to cure before things take a turn for the worse.

The damaging effects come from various compounds in this mushroom. There are at least eight toxins that have been discovered in the Death Cap. This mushroom targets the body's cells, affecting cell metabolism and leading to organ failure. Your liver and then your kidneys are in the most trouble, which is why so many people need a liver transplant after ingesting this mushroom. If you are unable to get a transplant in time, you may die.

Mortality Rate of the Death Cap

This mushroom has claimed many victims over the years. The mortality rate of the Death Cap was at 60-70 percent in the mid-twentieth century. Because of recent medical advancements the mortality rate has dropped to 10-15 percent, but that is not an invitation to try your luck with this mushroom.

Destroying Angel

Amanita bisporigera, A. ocreata, A. verna, and *A. virosa*

As indicated from the name, this is one group of mushrooms that foragers or mushroom-curious individuals should know to stay away from. Destroying Angel Mushrooms grow from June through November, and they will leave a white spore print when placed on a piece of paper. These mushrooms will have varied cap shapes and may dull with age. The Destroying Angel Mushrooms can be confused with some edible species such as:

+ the Button Mushroom
+ the Meadow Mushroom
+ the Horse Mushroom

In other words: Extreme caution is advised. Foraging should be left to the experts. When these mushrooms are in their egg-like phase, they may be confused for other edible mushroom "eggs," or for common edible puffball-like mushrooms. As for the specifics of their locations, *Amanita bisporigera* and *Amanita ocreata* are both found in North America, whereas *Amanita verna* and *Amanita virosa* are both found in Europe.

Careful with Your Pets

Destroying Angel Mushrooms are not just deadly for humans.
If you have a curious pup, or outdoor cat, verify the area
does not have any mushrooming bodies. Even though these
mushrooms are rare and you're not likely to find any, it's
worth taking precautions because just a bite or two of one
of them could kill. Exercise caution for your furry friends
and keep them leashed on unfamiliar land.

Destroying Angel Mushrooms and their cousin, the Death Cap Mushroom, are responsible for almost all the mushroom-related deaths around the world. These mushrooms inhibit your body's cells from properly functioning, and eventually lead to organ failure. An adult needs to eat only half a mushroom cap and symptoms will appear within 5–24 hours. The most common symptoms, which include vomiting, diarrhea, convulsions, delirium, and cramps, should be addressed at the emergency room as soon as possible. If mushroom poisoning is suspected, mycology experts will be consulted to identify the mushroom in question. The doctors might ask for a sample of the mushroom so they can get it positively identified. Unfortunately, though ingestion of these mushrooms is fairly common, there is no known antidote to the poison. Treatments may include IV fluids, blood filtering, and laxatives. Severe cases may need a liver transplant or large dose of penicillin. Compounds derived from the milk thistle plant are sometimes used to treat poisonings.

Fungi Theories

Proponents of a theory known as panspermia claim that
fungi and other basic life-forms were brought to earth via
meteorites. According to this theory, this is how life began
on this planet. It would have required strong and resilient
life-forms to not only survive that trip but thrive on their
new planet. Considering the harsh environments mushrooms
often live in, this may give some support to what many
consider to be an unsupportable theory.

DESTROYING ANGEL

DEVIL'S CIGAR

Devil's Cigar

Chorioactis geaster

AT A GLANCE

GEOGRAPHIC LOCATION ✦ Texas and Japan.

GROWING LOCATION ✦ Stumps or dead roots of cedar elms in Texas and on dead oaks and dead sapphireberry trees in Japan.

CHARACTERISTICS ✦ This mushroom starts cigar shaped, then splits open to reveal a four- to seven-point star-like arrangement.

PRIMARY USE ✦ N.A. (official Texas state mushroom).

The Devil's Cigar is also known as the Texas Star Mushroom or *Kirinomitake* in Japanese. *Kirinomitake* is named after the seed pods of the kiri tree. This mushroom is the only mushroom in the world to grow only in Texas and Japan and nowhere else. It's stumped researchers for many years; many have tried to figure out why this mushroom only shows up in two places separated by 6,800 miles, at approximately the same latitude.

It was thought that a person may have brought the spores from Texas to Japan or vice versa and they naturalized, but recent studies show that the two populations of mushrooms have been separated for over nineteen million years. This development has stumped scientists: Why has this happened nowhere else in the world? The mushrooms in Japan and Texas are slightly different, as the ones in Japan form from dead red-bark oak and dead Japanese sapphireberry trees, while in Texas they grow on the roots and stumps of cedar elms. The Japanese specimens have been grown in artificial culture in a lab, while the spores of the Texas specimens have not been able to be successfully germinated in culture in a lab. Despite their differences, and the great geographic distance separating them, the Texas and Japanese mushrooms look identical.

The first specimen was found in Austin, Texas, in 1893 by the American botanist and mycologist Lucien Marcus Underwood. Underwood sent them to mycologist Charles Horton Peck, who named them *Urnula geaster*. German-American botanist Elsie Kupfer named the fungi *Chorioactis geaster* in 1902.

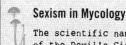

Sexism in Mycology

The scientific name change of the Devil's Cigar from *Urnula geaster* to *Chorioactis geaster* was debated for sixty-six years by various male mycologists, possibly because it was named by a female mycologist. Elsie Kupfer's choice has stood its ground and continues to be the valid Latin name.

The Devil's Cigar gets another one of its common names, the Texas Star, because it looks like a star and only grows in Texas. It was named the official Texas state mushroom in 2021. Texas tried to adopt this mushroom as the state species in 1997, but the bill failed to pass the state legislature. It gets its common name Devil's Cigar not only because it looks like a cigar before it splits open, but also because it goes through a process called dehiscence. Dehiscence is when the fruiting body splits open and shoots out a cloud of spores, resembling smoke from a cigar, which also makes an auditory hissing sound. This only happens in fifteen other mushrooms, so it's quite rare.

Outside Growth Patterns

In 2017, people reported finding one of these mushrooms in Oklahoma; it was the first time in nineteen million years this type of mushroom had been discovered outside of Japan or Texas.

Devil's Fingers

Clathrus archeri

Devil's Fingers, also called the Octopus Stinkhorn, is something out of a nightmare. It sprouts from an egg-like structure, also known as a peridium. The mushroom looks like a grouping of pinkish-red octopus tentacles from hell. Devil's Fingers has the additionally gruesome attribute of smelling like a pile of rotting meat. If found in nature, this mushroom will definitely have people wondering if they have entered a science fiction film, as it looks like an alien creature that just landed from space. Commonly found in gardens, this mushroom is known to cause fright for people who don't know what it is, and it is famous across social media around Halloween for its spooky qualities. The Devil's Fingers has an olive-brown spore print. The dark spots on the tentacle mushrooms are called gleba and are a slimy cluster of spores. The foul smell of these clusters attracts insects, and these insects transfer the spores to new locations. This all happens in the mushroom's growth season: from June to September. This species was first scientifically classified as *Lysurus archeri* by British mycologist Miles Joseph Berkeley in 1860 from a collection in Tasmania. Its name was changed to *Clathrus archeri* by British mycologist Donald Malcolm Dring in 1980.

As a saprophytic mushroom, the mycelium, or "roots," of the fungus secrete extracellular enzymes to break down organic matter for food. The mycelium starts as a white color but will eventually turn pinkish. On the surface of the mycelium, calcium oxalate crystals will form and then create a protective layer around the tips of the mycelium. In addition to enzymes used to break down organic matter, this fungus also secretes a type of acid that benefits the surrounding flora by increasing the bioavailability of minerals and changing the pH of the soil.

Since the compounds that Devil's Fingers produces smell almost the same as rotting flesh, they are appetizing for a lot of insects but not for humans. Unlike other stinkhorn species, the "egg" stage of this mushroom is not edible, nor is the mature mushroom. In other words, stay away from this mushroom.

Mushrooms Can Communicate with Their Hosts

Mycorrhizal fungi are able to communicate various amounts of information through their mycelium. Their mycelium intertwines with a plant's roots and can communicate information about oncoming dangers like lack of water, disease, or other such threats. In this way, the fungi keep their host safe, as well as other plants that share the same space as their host.

DEVIL'S FINGERS

DRYAD'S SADDLE

Dryad's Saddle

Cerioporus squamosus

Dryad's Saddle, also known as Pheasant's Back Mushroom, was first discovered in 1778 by British botanist William Hudson. Hudson originally called it *Boletus squamosus*. The name was changed by Swedish mycologist Elias Magnus Fries to *Polyporus squamosus* in 1821. However, it was changed once again in 1886 to its current Latin name by French mycologist Lucien Quélet. Although the Latin name was changed almost two hundred years ago, a lot of books and people still refer to this mushroom as *Polyporus squamosus*, so keep in mind that they are actually referencing *Cerioporus squamosus*.

Dryad's Saddle is one of the first mushrooms you are likely to find in the springtime if you are out looking for Morels and other early spring mushrooms. Although this mushroom is edible, it is not very tasty and only truly edible when it's young. As it grows older and larger, it hardens and gets rubbery, making it almost inedible. If you find an older specimen and want to eat it, these mushrooms can sometimes be cut around the leading edge of the mushroom to get the newest growth. But most of the time, by this stage the mushrooms are filled with maggots and waterlogged. There's only a small window during which these mushrooms are at their prime. When cooked, their flavor is often compared to a nutty rind of a watermelon. If that sounds appealing to you, keep an eye out for the new growth. In general, Dryad's Saddle is mostly left unpicked, except for the brave few who rely on wild edibles as their main food source or know how to get prime specimens and cook them properly. A good indicator if this mushroom is good to eat is the color of its stem. As Dryad's Saddle ages, the stem turns from white to black. When it's young and good to eat, the stem is white; once the stem has darkened in color, the mushroom is no longer considered safe to consume.

There is little research on this mushroom's health benefits, but it has been used for antioxidant support and immune support. More research will need to take place before it is truly recommended as a functional mushroom. An interesting use of this mushroom is for papermaking. Some individuals will hunt Dryad's Saddle specifically for this use. The stiff paper turns out to have an almost skin-like, translucent quality that is yellowish and tan in color with speckles throughout.

A Greek-Inspired Name

The name Dryad's Saddle comes from Greek mythology. In the realm of Greek myths, dryads, also called tree nymphs, had magical, nature-related powers. These creatures would use these mushrooms as a form of transportation—sitting and riding on them from one place to another.

Earthstars

Geastraceae family

AT A GLANCE

GEOGRAPHIC LOCATION ✦ Global.

GROWING LOCATION ✦
The forest floor.

CHARACTERISTICS ✦ Generally,
a puffball-like spore sac
surrounded by star-like petals.

PRIMARY USE ✦ Some species
are functional.

This mushroom family's Latin name comes from *geo* meaning "earth" and *aster* meaning "star." Earthstars are known for their odd appearance, almost like a starfish and flower combined. These decomposers are compared to Puffball Mushrooms, as well, but they have some key differences. When they are young, they appear almost like Puffballs with beaks. However, Earthstars are unique from Puffball Mushrooms because they have two peridiums, or protective layers, protecting the spores instead of the Puffball Mushroom's one. The inner peridium is just like a Puffball Mushroom, where it looks like a circular sac of spores, but the outer peridium eventually lowers and makes a star shape of fungal petals giving rise to the name Earthstars.

The family Geastraceae, in the order Geastrales, holds eight genera. The mushrooms in these eight genera can all arguably be considered Earthstar Fungi: *Geasteroides, Geastrum, Myriostoma, Nidulariopsis, Phialastrum, Radiigera, Schenella,* and *Sphaerobolus*. The Earthstars are a diverse lot:

+ *Geasteroides texensis* is the only species in the genera *Gaesteroides* and is a Texas native Earthstar that grows around stumps of post oak trees.

Mushroom Tumbleweeds?

When they are dried up, some Earthstar Mushrooms have the ability to have their petals curl around the spore sac to protect it. In these cases, the mushroom can then roll around like a tumbleweed. When it gets humid and starts raining, the star-like petals will open up and when a raindrop hits the spores, it will launch its DNA into the forest.

+ *Geastrum* houses more than a hundred species of Earthstar fungi with probably the most famous being *Geastrum saccatum*, also called the Rounded Earthstar and known for its organic compounds used to support the immune system, provide key antioxidants, and also support the body's natural inflammatory response post workout.

+ *Myriostoma* is a genus that comprises five species that are commonly called Salt Shaker or Pepper Pot Fungi because they have multiple holes in the spore sac, which makes it look like a salt or pepper shaker. When a raindrop hits this mushroom, it creates more of a dispersed cloud of smoke instead of the usual singular puffed burst.

+ *Nidulariopsis* is a genus that comprises two species that are incredibly understudied. These mushrooms need more research to determine if they have potential uses.

+ *Phialastrum* is a genus that has only one species, *Phialastrum barbatum*, which was found in the Democratic Republic of the Congo.

+ *Radiigera* contains only four species: *R. bushnellii*, *R. flexuosa*, *R. fuscogleba*, and *R. taylorii*. These species also need more research to see if they have any benefits.

+ *Schenella* also contains only four species, *S. microspora*, *S. pityophila*, *S. romana*, and *S. simplex*, which are scattered around Africa, North America, and Europe. These species also require more research.

+ *Sphaerobolus* is a genus that gets tagged with the common names Shotgun Fungus, Artillery Fungus, or Cannonball Fungus, as they blast out the spores with explosive velocity.

EARTHSTARS

ENOKI

Enoki

Flammulina velutipes

Enoki, also known as Velvet Foot, the Winter Mushroom, or Velvet Shank, is a bright orange mushroom that grows in clusters on hardwoods. They can be found growing either on an upright hardwood tree, such as elm, ash, beech, or oak, or on fallen wood that's on or buried in the ground.

Until 2018, the Enoki's scientific name, *Flammulina velutipes*, was also given to one of the most cultivated gourmet mushrooms in the world, Enokitake, which is extremely popular in Japan. Mushroom enthusiasts cultivate Enokitake in bottles in the dark, where they grow incredibly long and skinny stems and very tiny caps. These bottle-grown mushrooms completely lose their color and turn pure white. In the wild, if part of a mushroom cluster grows in between the tree bark, the mushrooms will also look exactly like this: long skinny stems, small caps, and pure white in color. In 2018, DNA analysis revealed that the bottle-grown cultivated variety, which was thought to be *Flammulina velutipes*, was actually a different species. This new Enokitake species was given the Latin name *Flammulina filiformis*; this is one of the best mushrooms to have in soups. *Flammulina filiformis* has actually been cultivated since 800 C.E.; between China and Japan, they produce millions of tons per year to supply the culinary market.

Mushroom Growth in Outer Space

Cultures of the Enoki Mushroom went to space in 1993 when they were grown on the space shuttle *Columbia*. Researchers wanted to see how mushrooms would grow in a low-gravity environment. In the wild, these mushrooms usually orient their gills to face toward the ground for optimal spore dispersal, but in this space experiment the mushrooms had no orientation because of the lack of gravity and were growing in all different directions.

Now that it has been revealed that Enoki Mushrooms, or *Flammulina velutipes*, are not the same as Enokitake, or *Flammulina filiformis*, we will see if people start artificially cultivating Enoki at scale as well. After all, this mushroom is also a highly sought-after wild mushroom and would potentially grow very much like Enokitake. Like any mushroom, it's important to have positive identification before you eat it. However, take special care with Enoki. This mushroom looks like two very deadly and poisonous mushrooms, both in the *Galerina* genus.

Enoki Recipes

Enoki Mushrooms are delicious when cooked. They have a nutty flavor with a crisp, yet chewy, texture. They are very popular in Asia, where they are served in noodle dishes and hot pots. In Japan, nametake is a popular condiment that has a combination of simmered Enoki and other traditional Japanese flavors, including soy sauce and mirin.

Fly Agaric

Amanita muscaria

Fly Agaric, also known as Fly Amanita, has the quintessential mushroom "look." Getting recognition on postcards and T-shirts, in video games, as the houses in the Smurfs franchise, and even as the standard mushroom emoji on almost every technological platform, this mushroom is known for its bright colors and generic cap-and-stem shape. For people who grew up playing Super Mario video games, the mushroom that gives Mario his size boost and is one of the most prevalent icons of this hugely popular game series, looks just like a Fly Agaric.

There are theories that this mushroom was responsible for the Santa Claus story most people know today. The Indigenous people of Siberia (the Sámi) have been using this mushroom for ages. They grow underneath evergreen trees, like presents under a Christmas tree, and once picked they are either hung on the tree to dry, like ornaments, or stuffed into a stocking over a fireplace. Sometimes the Sámi chop off the caps, which they call "cookies," and weave them together to make a wreath. The story goes that around the winter solstice, a shaman would go from hut to hut on a sleigh pulled by reindeer with a bag full of mushrooms and give them to people as gifts. The reindeer would like to eat these mushrooms as well, so they might have felt like they were flying, and the shaman eating them was jolly, just as the stories describe. A lot of times there would be so much snow that the mushroom Santa Claus shaman would have to climb into the smoke hole or chimney with his bag of gifts. There was no currency exchange, only exchange of food, just like leaving out cookies, carrots, and milk for Santa Claus and the reindeer. The dried mushrooms were also rehydrated in milk to eat as a snack. Whether or not this is the true origin story of Santa Claus, we will never know, but it makes for great trivia to share with your friends.

The Fly Agaric is semi-toxic because it contains a toxic acid called ibotenic. However, this acid gets broken down in a person's liver into a psychedelic compound called muscimol. Those looking to cook with Fly Agaric should boil mushrooms and keep removing and replacing the water to remove the toxic ibotenic acid and psychedelic muscimol to make a nonpsychoactive and edible meal. If you wanted a psychoactive experience, you would have to use dry heat to convert the acid into its psychedelic compound. However, the psychoactive effects of the Fly Agaric are considered very unpredictable; it is not a popular mushroom choice of those looking for a psychoactive experience.

How the Fly Agaric Got Its Name

Fly Agaric traditionally was used to repel flies. When torn up and thrown into milk, the mushroom's acid would become a potent fly magnet. After luring the flies, the acid ultimately kills them. It was the insecticide of olden times.

FLY AGARIC

FROST'S BOLETE

Frost's Bolete

Butyriboletus frostii

The Frost's Bolete, also known as the Candy Apple Bolete or the Apple Bolete, is a red mushroom that turns blue when bruised. The changing-color reaction can be a cool trick to show people unfamiliar with this mushroom on a hike. Take a mushroom and carefully cut it in half. Then, watch the mushroom quickly change to blue before your eyes.

This mushroom was originally named *Boletus frostii* in 1874 by American minister John Lewis Russell. Russell named Bolete's Frost after his friend, the American mycologist Charles Christopher Frost. Frost returned the favor, naming the *Boletus russellii* mushroom after Russell. In science, when a species is named after a person, the naming must be done by someone else. However, it was found out later that Charles Christopher Frost actually named both mushrooms. Frost got his friend John Lewis Russell to publish his findings so he could name it *Boletus frostii*. In return, Frost named a mushroom after Russell. In this way, the clever Frost was able to honor both himself and his friend.

Because of his tactics, Frost's last name lives on in both the common name and the current Latin name of this mushroom.

Many Name Changes of the Frost's Bolete

Since this mushroom was originally described, it has gone through several name changes. First, it was named *Suillellus frostii* in 1909 by American mycologist William Murrill, then *Tubiporus frostii* in 1968 by Japanese mycologist Sanshi Imai, then in 2014 it was changed to *Exsudoporus frostii* by Simonini and Gelardi Vizzini, and then finally it was given the current Latin name, *Butyriboletus frostii*, by mycologists G. Wu, Kuan Zhao, and Zhu L. Yang in 2016.

Another common name for the Frost's Bolete Mushroom is Apple Bolete. The reason for this is because if the mushroom dries out after a rainstorm, the cap becomes especially shiny and red like an apple. In Mexico it's called *Panza Agria*, which translates to "sour belly," potentially because of its acidic citrusy flavor.

Although Frost's Bolete is edible and commonly sold in markets in Mexico, some people don't have the best gastrointestinal reaction to it. In other words: If you're going to eat this mushroom, proceed with caution. Also, unless you have this mushroom verified by a professional, don't eat it. It can be confused with two other red-capped boletes: *Boletus flammans* and *Boletus rubroflammeus*. Both mushrooms, while not deadly, are poisonous. Another reason to be careful is that Frost's Bolete can be infected and parasitized by a mold-like fungus called *Sepedonium ampullosporum* that causes the tissue to die and turn yellow.

Lightning Affects Mushroom Growth

Mushrooms have so many hidden surprises, is it really a wonder that some species grow better when lightning strikes? In Japan, lightning was always thought to help with mushroom growth. According to researchers, some mushrooms can double their output when lightning strikes in their vicinity.

Glowing Panellus

≼ Panellus stipticus

AT A GLANCE

GEOGRAPHIC LOCATION ✦ North and Central America, Europe, Asia, Australia, and New Zealand.

GROWING LOCATION ✦ Saprophytic, growing mainly on hardwood trees, logs, stumps, and fallen branches. This mushroom has been found to grow on pine as well.

CHARACTERISTICS ✦ White to tan in color, they have kidney- or clamshell-shaped fruiting bodies that grow in clusters with gills on the underside. In the dark, the mushrooms and mycelium glow a bioluminescent greenish color.

PRIMARY USE ✦ Functional and mycoremediation; technically edible but ingesting them is not recommended due to bitter taste.

Glowing Panellus, also called Bitter Oyster, Astringent Panus, Luminescent Panellus, or the Styptic Fungus, is most famous for its ability to glow in the dark. Its bioluminescence is caused by enzymes which degrade a pigment, emitting light. As of the time of writing this book, there are one hundred and five species of glow-in-the-dark fungi, with more being discovered annually. Each bioluminescent fungus glows in different areas, whether the mycelium, the gills, the stem, the cap, the whole mushroom, or a combination of the aforementioned parts. The term *foxfire* was given to glowing fungi found in the wild.

The intensity of the glow from *Panellus stipticus* is low compared to very bright organisms like a firefly, but they can continue glowing for days to months on end making the total amount of "glow emissions" pretty high. These fungi are thought to have developed luminescent capabilities to attract insects at night so that the insects would transfer their spores farther in the forest. For *Panellus stipticus*, the fruiting bodies are much brighter than the mycelium, but all parts of the fungus glow. However, strangely enough, even though this mushroom grows all over the world, it has been reported to glow only in the east coast of North America. The reason why it glows in this region and not in others is currently unknown.

Glowing Panellus was first named *Agaricus stipticus* in 1783 by French botanist Jean Bulliard. The name was then changed by a large number of other scientists. *Panellus stipticus* became the mushroom's current name back in 1879 when named by Petter Karsten, mycologist from Finland.

Although the Glowing Panellus is technically edible and non-poisonous, eating it is not advisable because of its small size and bitter taste. Historically this mushroom has been used in traditional Chinese herbalism to stop bleeding if dried, powdered, and put into open wounds. In scientific research, it has been noted that certain biological pollutants lower the bioluminescence from this mushroom, so researchers are considering using it for bioindicators of pollutants, like a canary in a coal mine. Not only can it be used to detect pollutants in the environment, but it can also be used in the process of mycoremediation for degrading pollutants in wastewater.

Natural Night-Lights

Grow kits are sold online that allow you to grow the mycelium or even the mushrooms of *Panellus stipticus* and have them as biological night-lights in your house. The glow is very subtle indoors or outdoors, but once your eyes get adjusted it's a very cool glowing green color.

GLOWING PANELLUS

GOLDEN THREAD CORDYCEPS

Golden Thread Cordyceps

Tolypocladium ophioglossoides

The Golden Thread Cordyceps is a club-like mushroom with a bumpy head. This mushroom is interesting in that it grows on a false truffle, commonly known as deer truffles, in the *Elaphomyces* genus. *Elaphomyces* false truffles grow in temperate and subarctic regions in symbiosis with primarily spruce, pine, and beech trees. These *Elaphomyces* fungi trade nutrients with their tree hosts. Therefore, when looking for the Golden Thread Cordyceps you will always find the mushroom in relation to a tree. Most likely if you find one, you will find a colony in relation to the roots of the same tree or tree grove. It's easy to mistake this mushroom for an Earth Tongue Mushroom, Dead Man's Fingers, or other *Cordyceps or Ophiocordyceps* species. However, this mushroom does have a dead giveaway: the defining feature of a golden thread leading down to the deer truffle.

When harvesting this mushroom, it's important to carefully dig around the base of the mushroom, excavate for it, and follow the thin fragile golden mycelial thread to locate the deer truffle. When getting this mushroom, it's important to collect the mushroom, thread, and truffle all attached together. Some researchers speculate that over seventy million years ago this mushroom used to infect insects like some *Cordyceps* species, but it switched to parasitizing fungi instead. The theory is that both insects and these truffles live in similar habitats, which made the host-jumping event easier.

There are a few main functional compounds which have been isolated from this fungus within the fruiting body and in the mycelium. Functionally, this helpful mushroom can be used to support the immune system. The Golden Thread Cordyceps is also used in traditional Chinese herbalism to support lung and kidney health and energy levels.

Traditional preparations of Golden Thread Cordyceps involve stuffing it into a duck and making a broth. However, most guides indicate that both the deer truffle and the mushroom are inedible, so proceed with caution. If eaten, the mushroom's pale interior has a mild and musky flavor. This mushroom can be grown on rice in a similar way to *Cordyceps militaris*. The author successfully grew some Golden Thread Cordyceps at his farm in Massachusetts on a nutrient rice medium in a Mason jar.

Historic Mushroom Use

There is evidence that humans were eating mushrooms as far back as 17,000 B.C.E. The skeleton of the "Red Lady of El Mirón" was found in Spain with two types of mushroom spores in her teeth. After this, in 4713 B.C.E. there are indications that the Tassili people from Algeria used mushrooms in rituals. Then, in 4000 B.C.E., the Selva Pascuala was created. This Spanish cave mural depicted many mushrooms.

Hedgehog Mushroom

Hydnum repandum

The Hedgehog Mushroom, also called the Sweet Tooth Mushroom, is a choice edible species from Europe and North America that is beloved by most everyone who consumes it. This mushroom was first described in 1753 by Swedish botanist Carl Linnaeus, sanctioned as *Hydnum repandum* in 1821 by Swedish mycologist Elias Magnus Fries, but became official in 1977 after a nomenclatural proposal by American mycologist Ronald H. Petersen. DNA analysis done in 2009 and 2016 suggests that the umbrella name *Hydnum repandum* houses at least four separate mushrooms across North America, Europe, and Asia. It's also likely that with new DNA analysis, these mushrooms will get new Latin names in the future, so stay tuned.

This mushroom has no poisonous look-alikes, and it can be easily identified by beginners, which makes it a prime edible species for beginning foragers. Two similar common species in North America are *Hydnum albidum* (White Hedgehog) and *Hydnum albomagnum* (Giant Hedgehog), which are both great edible species. For a *true* beginner, this mushroom could potentially be mistaken for Chanterelles, which could then be mistaken for the Jack O'Lantern

Mushrooms. If the Jack O'Lantern Mushrooms are eaten, they will give you an upset stomach, vomiting, and diarrhea. The easiest way to tell a Hedgehog Mushroom from a Jack O'Lantern Mushroom is by looking beneath the cap. The recognizable teeth on the *Hydnum* species will be visible, instead of the gills on Chanterelles and the Jack O'Lantern Mushrooms.

The Hedgehog Mushroom is a prized edible species. It's considered to be on the same tier as Chanterelles, which are world-renowned and highly sought after. The Hedgehog Mushroom's taste is described as sweet, fruity, and nutty with a refreshing and crunchy texture. What's even better is that it's unlikely to be infested with maggots, as opposed to other highly sought-after edible mushrooms like the Porcini, where this is a common theme. The only pest you need to watch out for is squirrels, especially the red squirrel, who also love eating this mushroom.

The Hedgehog Mushroom also holds up well in the freezer, so if you don't eat it all, you can save it for later. This is a plus, as not all mushrooms freeze well. The culinary adventures you can have with this mushroom are virtually endless: Try sautéing, pickling, roasting, baking, and even simmering this mushroom in milk, stews, or stocks. Like most mushrooms, it absorbs any flavor you are cooking with, making it a great base for any dish. Two things to be mindful of when picking this mushroom:

+ It easily absorbs radioactive isotopes (making it a poor choice to pick around Chernobyl).
+ It was put on the European Red List as potentially endangered. Note: On the list, it is rated as of "least concern."

Popularity in the Western Hemisphere

The Hedgehog Mushroom is commonly collected and sold in markets for food in Mexico, Canada, the United States, and all over Europe, including Spain, France, Italy, and Finland. This mushroom's delicious flavor is highly respected and coveted.

HEDGEHOG MUSHROOM

HONEY MUSHROOM

Honey Mushroom

 Armillaria spp.

AT A GLANCE

GEOGRAPHIC LOCATION ✦ Global.

GROWING LOCATION ✦ Parasitizes a wide range of tree and shrub hosts.

CHARACTERISTICS ✦ Cap-and-stem mushrooms that have colors ranging from white to golden. Some mushrooms in this genus are bioluminescent. The mycelium creates thick black cords which may look like thin vines. These "vines" take over living trees.

PRIMARY USE ✦ Edible.

Honey Mushrooms, also known as Honey Fungus, are a group of about forty species of parasitic mushrooms. These mushrooms are found around the world and parasitize primarily living trees and shrubs, but can also be found on dying, or already dead, woody material as well.

Armillaria was first described in 1821 by famous Swedish mycologist Elias Magnus Fries. These mushrooms were then assigned generic rank by German mycologist Friedrich Staude in 1857. Finally, American mycologists Tom Volk and Harold Burdsall cleaned up the genus in 1995 to determine the forty species that are true *Armillaria* spp.

This group of fungi is most famous for the species *Armillaria ostoyae*, one example of which is growing in Oregon and considered to be the largest organism in the world. This fungus in the Malheur National Forest is estimated to be around 2,500–4,000 years old and covers about 2,200 acres. Scientists used DNA analysis to test various sites across this large area and determined that this giant fungal organism is all connected underground; it is one giant fungal network. This mushroom network has been able to grow as big as it is because the environment is old and undisturbed. Although some scientists would classify this as a parasitic fungus that is causing environmental damage to large swathes of forests, other ecologists refer to this fungus as a "meadow maker." The latter group claims that the fungus is crucial for ecosystems to create patches of pasture for other organisms to thrive.

Not All Honey Mushrooms Are Parasites

Although most species of Honey Mushroom are parasitic and saprophytic (living on dead organic matter), there are some species which form mycorrhizal connections with orchids, meaning that the mycelium of this fungus connects with the roots of orchids to trade nutrients. Orchids rely on mycorrhizal fungi for the first stages of their life as, when in the seed stage, they don't have any energy reserves and rely on the fungi entirely for a source of carbon.

Researchers are currently looking into the biotechnological potential of these mushrooms. Honey Mushrooms produce electric signals when attached to extracellular electrodes, meaning they could be used in bio batteries in the future, thus creating more eco-friendly power sources. Another cool fact is that this specific fungus is bioluminescent, and the mushrooms will glow in the dark. Other species of *Armillaria* are also bioluminescent, including *Armillaria gallica*, *Armillaria mellea*, and *Armillaria tabescens*.

Honey Mushrooms are delicious edible species across the world. As usual, when eating mushrooms, it's important to exercise caution. Some mycologists suggest avoiding alcohol for twelve hours before and twenty-four hours after consuming these mushrooms to avoid gastrointestinal issues. As with most mushrooms, it's best to cook this mushroom; it can be mildly poisonous when eaten raw.

Indigo Milky Cap

Lactarius indigo

Indigo Milky Cap, also called Indigo Milky or Blue Milk Mushroom, is a prized edible mushroom species. In 1822, German-American mycologist Lewis David de Schweinitz first described this mushroom. Then, in 1838, Swedish mycologist Elias Magnus Fries gave the Indigo Milky Cap its official, recognized Latin name, *Lactarius indigo*. Though other mycologists attempted to recategorize the mushroom in 1960, its name given by Fries remained.

As a mycorrhizal fungus, this mushroom forms a symbiotic relationship with the roots of deciduous and coniferous trees. Therefore, it will always be found on the ground near those trees. The mushrooms secrete enzymes that mineralize organic compounds, then trade through the roots of the tree for sugars from the tree. Although it's primarily found in North or Central America and East Asia, there are reports of Indigo Milky Cap being found in the south of France, indicating that it may pop up in other places as well.

Blue from its stem to its gills to its cap, this is a gorgeous mushroom to set eyes on. It even secretes a bluish latex milk if cut or damaged, leading to its name. The bright indigo-blue color makes this mushroom stand out in the forest; even if you don't eat it, it's still a pretty mushroom to experience. The blue color comes from a specific molecule that is similar to azulene. Because of its color, the blue pigment of Indigo Milky Caps has been used as a basis for man-made fluorescent pigments.

Unfortunately, the Indigo Milky Cap's blue color gets cooked out in a frying pan and turns to a grayish color. In Mexico, this mushroom is used as a topping for tacos on blue corn tortillas for a beautiful blue-themed dish. The mushroom is routinely sold in markets in Mexico, Guatemala, and China. Some people even use this mushroom to color marinades in Canada. Though it seems as though it is used widely in cooking, people disagree about how tasty it is: Some call it mediocre, at best. This might have to do with the fact that not many people know how to cook *Lactarius* mushrooms. They are often gritty in texture unless you do a wet sauté first to break down the cell walls to enhance the texture, and then an oil sauté after to get a nice golden-brown color. This is the secret to turning mediocre *Lactarius* mushrooms into delicious treats. Sadly, the Indigo Milky Cap doesn't dry well, so it's sold fresh. No one has figured out how to artificially cultivate this species, so it can only be foraged in the wild.

Mushroom Party Trick

If you happen across this mushroom while taking a walk outside with friends, the Indigo Milky Cap can make for a cool party trick. Simply cut it open and see the white oozing milk or latex substance come out. This mushroom is quite a sight for someone who isn't familiar with it, and you can show them what makes it so famous.

INDIGO MILKY CAP

JACK O'LANTERN MUSHROOMS

Jack O'Lantern Mushrooms

Omphalotus illudens, O. olearius, O. olivascens, and *O. subilludens*

AT A GLANCE

GEOGRAPHIC LOCATION ✦ North America and Europe.

GROWING LOCATION ✦ Decaying stumps, roots, or at the base of hardwood trees.

CHARACTERISTICS ✦ Cluster of bright orange cap-and-stem mushrooms with gills underneath.

PRIMARY USE ✦ N.A. (poisonous but potentially functional).

Jack O'Lantern Mushrooms are a well-known group of mushrooms in the *Omphalotus* genus. This genus includes *Omphalotus illudens, O. olearius, O. olivascens,* and *O. subilludens.* These four mushrooms all are bright orange cap-and-stem mushrooms with gills underneath the caps. They grow in clusters on wood either from the base of a tree or from fallen wood on, or beneath, the ground. Jack O'Lantern Mushrooms are found in North America and Europe.

All *Omphalotus* species contain two different types of toxins that, though not lethal, can possibly damage a human's DNA and cells. The toxins can also cause cramps, vomiting, and diarrhea within an hour of eating the mushroom. The symptoms will last around 1–3 days. It's unclear if the toxins will cause long-term damage on cells and DNA, as the mechanisms of these toxins are not yet well understood. The toxins are currently being studied for their use fighting some types of cancer, or as an antifungal or antibiotic. However, the toxins would need to be altered as they are too damaging to the human body. MGI Pharma, a cancer-focused pharmaceutical company, created a drug called Irofulven, which is an analog of one of the toxins from the Jack O'Lantern Mushrooms and other species, as an experimental cancer drug which is now going through clinical trials.

Look-alikes can include the highly sought-after Chicken of the Woods, Chanterelle species including *Cantharellus lateritius* and *Cantharellus cibarius*, Honey Mushroom varieties including *Armillaria mellea* and *Desarmillaria caespitosa*, and psychedelic species like the Big Laughing Gym (*Gymnopilus junonius*). The one obvious indicator when foraging at night is that *Omphalotus illudens*, *O. olearius*, *O. olivascens*, and *O. subilludens* all glow in the dark from their gills. The bioluminescence comes from an enzyme that degrades a compound, which leads to the emission of light.

Similar species in the *Omphalotus* genus include:

- *Omphalotus flagelliformis* in China
- *Omphalotus guepiniformis* in Russia
- *Omphalotus japonicus* in Korea, China, Japan, and Russia
- *Omphalotus manensis* in China
- *Omphalotus mexicanus* in Mexico
- *Omphalotus nidiformis* in Australia and India

Although these other species in the genus lack the common name Jack O'Lantern Mushrooms, they are all bioluminescent and have the same toxic compounds, so they could absolutely be lumped under the same Jack O'Lantern umbrella.

Glowing Like Jack O'Lanterns on Halloween

These mushrooms will continue to glow for 24–48 hours after they are picked. They can act as natural lanterns, of sorts. Picking and carrying these mushrooms around can be a fun activity for kids. You might want to seal these mushrooms in child-proof, clear containers, or warn your kids about the toxicity of these mushrooms. If ingested, no one will be having a good time.

Jelly Babies

Leotia lubrica

Jelly Babies are cute, small (0.4–2.4 inch) mushrooms. They look very similar to the popular Japanese "Kinoko no Yama" treats from the company Meiji, which are tiny mushroom chocolates with a biscuit stem.

The texture of Jelly Babies can be slimy, clammy, or smooth. Additionally, the cap of the mushroom tends to vary in shape, embodying unique irregular up-and-down outlines, rolls, and lobes, curling inward toward the stem. The name *Leotia*, meaning "slimy," was given because of the slimy caps, and *lubrica* means "slippery" or "smooth." This mushroom was originally named by Italian naturalist Giovanni Antonio Scopoli in 1772, but he couldn't decide between the names *Elvella lubrica* and *Helvella lubrica*. In 1794, German mycologist Christiaan Hendrik Persoon gave these mushrooms the name *Leotia lubrica* and the name hasn't changed since.

Jelly Babies love growing in soil, moss, and plant wastes usually in damp, mainly deciduous forests, but sometimes under conifers. They most commonly grow in clusters, but every once in a while, you can find a solo specimen. Similar species that are often confused with *Leotia lubrica* include:

- ✦ *Cudonia confusa* (also known as the Cinnamon Jelly Baby)
- ✦ *Cudonia circinans*
- ✦ *Cudonia lutea*
- ✦ *Leotia viscosa* (also known as the Green Jelly Baby)
- ✦ *Leotia atrovirens*
- ✦ *Leotia viscosa*

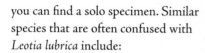

Mushroom of Many Names

Perhaps it's because of the adorable nature of these mushrooms, but the Jelly Babies have a variety of other cutesy nicknames throughout the world. Other fun names for this mushroom include Slippery Cap, Gumdrop Fungus, Green Slime Fungus, Ochre Jelly Club, and Lizard Tuft.

Although regarded as saprophytic, meaning the fungus decomposes dying or dead organic matter, there is some evidence that Jelly Babies could be ectomycorrhizal. Research is still being done regarding this claim, and evidence is inconclusive at this point. When finding these mushrooms in the wild, there is a chance a parasitic fungus, *Hypomyces leotiarum* (also known as the asexual stage of *Hypomyces leotiicola*), may be attacking the Jelly Babies, causing discoloration to their caps.

Jelly Babies are not poisonous—which is likely for the best, given their appetizing name. However, most mycologists and guidebooks list this mushroom as inedible. This is probably because of these mushrooms' tiny size and slimy consistency. Other mycologists consider these mushrooms edible, just bland, and Charles McIlvaine (a famous American mycologist) even considered these mushrooms as tasty snacks. So, consider the edibility status of these mushrooms a little up in the air—they're most likely edible but considered undesirable by most.

Mushrooms Produce Vitamin D

Mushrooms contain ergosterol, which is a substance that can convert to vitamin D. When exposed to the sun, mushrooms can convert this provitamin into the more useful vitamin D. This means that mushrooms, on their own, are a source of vitamin D.

JELLY BABIES

LION'S MANE

Lion's Mane

 Hericium erinaceus

*L*ion's Mane is one of the most well-known functional mushrooms in the world. It grows on hardwood logs, but occasionally it can be found growing on a still-living tree. The white group of tendrils gives it its famous name: Lion's Mane. This mushroom is not only great for health and wellness, but it is also an amazingly tasty culinary mushroom that you can find at grocery stores and farmers' markets.

Other names for this mushroom are Pom Pom Mushroom, Monkey Head Mushroom, *Hou You Gu* in China, and *Yamabushi-take* in Japan. Lion's Mane can easily be found in the wild, but it is also widely commercially cultivated either indoors in bags filled with hardwood-supplemented sawdust or outdoors on logs. It is one mushroom species that is a gift that keeps on giving because it will continue to grow on a tree for up to twenty years. Lion's Mane can also grow from a grow bag up to eight times without needing a break.

The taste of this mushroom has been described as similar to crab, lobster, or fish. You can use Lion's Mane in many recipes. For example: Cut it up in strips, use an egg wash, bread it, and cook to a nice golden brown on both sides to make "Lion's Mane breaded fish."

Lion's Mane is referred to as the "smart mushroom" or "brain mushroom" because of its supportive benefits for the brain. It is used to support cognitive function, memory, focus, as well as the digestive system and immune system. This mushroom is also considered an adaptogen, meaning it helps support the body's ability to deal with occasional stress and fatigue. Some people like to take Lion's Mane (or mushroom supplements) before bed to aid in their sleep and to help their brain "recharge" while they are sleeping. Other people like to take Lion's Mane for work or studying.

Like all functional mushrooms, Lion's Mane is packed with amazing compounds. Many of the compounds within Lion's Mane support the natural production of nerve growth factor in the brain. Nerve growth factor is important in the development, survival, and proliferation of nerve cells (neurons), which have to do with transmitting vital information in the body. Like all functional mushrooms, it's important to use a dual extraction to get both polar and nonpolar compounds inside. With both types of compounds, you will best support your overall health and wellness. Because the main compounds in Lion's Mane are nonpolar, it is vital to use alcohol to extract the compounds to get those brain-supportive goodies.

Other Ways to Eat Lion's Mane

This mushroom is diverse in its uses in the kitchen. As it has a nice, umami flavor, you can easily put it in an omelet or with other egg dishes (like a quiche). You can toss it in your favorite pasta dish or throw it into a curry. Some people will even find it appetizing to just lightly pan-fry with some garlic. Lion's Mane Mushroom has limitless possibilities.

Lobster Mushroom

Hypomyces lactifluorum

Lobster Mushroom is actually a parasitic fungus that attacks Milk Caps and *Russula* mushrooms. The mold, being a burnt orange to bright red color, makes the deformed host mushroom look like a lobster tail or shell. Hence this fungus's common name: Lobster Mushroom.

When the fungus attaches itself to a host species, it changes the chemistry of the host mushroom. The fungus then transforms its host into a whole new organism that becomes highly desirable as a delicious edible mushroom. *Lactarius piperatus*, a common host of the Lobster Mushroom fungus, is ordinarily hot and spicy in flavor, but when it's colonized, its chemistry changes, making the new parasitized version taste really good.

A study done in Quebec showed that over time *Hypomyces lactifluorum* took over its host, *Russula brevipes*, so thoroughly that DNA tests showed only trace amounts of the *Russula*. This means that the majority of the DNA of the newly transformed fungus was actually the *Hypomyces* parasite. Not only that, but the study showed that over the course of the infection, the Lobster Mushroom fungus completely changed the chemical makeup, the diversity, and the number of metabolites produced by the *Russula* species.

Although this Lobster Mushroom fungus is quite unique and there are not very easily mistaken look-alikes that are poisonous, one should always know what they are looking for before picking any mushrooms with mold growing on them for consumption.

The taste is considered to be a slightly seafood-like flavor that some chefs say is best served alone. The texture is crunchier than other mushrooms. Most chefs will prepare this Lobster Mushroom in similar ways they would prepare any seafood, like in pasta, egg dishes, soups, chowders, cream-based sauces, stews, terrines, risotto, and stir-fries. This fungus pairs really well with cauliflower, broccoli, spinach, orzo, potatoes, garlic, ginger, lemon, tomatoes, and asparagus. When cooking, the orange or red color will leach out, and can be used for an interesting sauce, juice, or dressing.

Not only can the color separated from the mushroom be used for cooking, but it can also be used for dyeing fabrics or paper. The colors you can get from the Lobster Mushroom include a pale orange to pink, saffron, shades of purple, red, and more. You can make an interesting postcard or gift during the holidays. People might give you a second look when you say you dyed your handwritten card with mushrooms. You can even go a step further and make mushroom paper with another mushroom like Turkey Tail or Reishi and then dye it with Lobster Mushrooms.

How to Tell When Lobster Mushrooms Go Bad

As with many other mushrooms, Lobster Mushrooms give off particular signs when they are no longer fresh or safe to eat. When they rot, they produce a terrible smell. They may lose their bright orange color, and they may go slimy to the touch. If they show any of these signs, or if they change texture, your mushrooms have expired.

LOBSTER MUSHROOM

MAGIC MUSHROOMS

Magic Mushrooms

Psilocybe cubensis (and others)

Magic Mushrooms, also called Shrooms, are a common title given to *Psilocybe cubensis*, as it's the most commonly cultivated and eaten psychoactive, or "psychedelic," mushroom in the world by far. However, the term *Magic Mushrooms* may refer to any mushroom containing the psilocin or psilocybin molecules, which are famous for giving the hallucinatory effect. The Fly Agaric Mushroom could also potentially be given the title of Magic Mushroom, as it also produces hallucinogenic effects, but the title is more commonly given to mushrooms that contain psilocybin/psilocin.

Psilocybe cubensis was first called *Stropharia cubensis* by American mycologist Franklin Sumner Earle in 1906. This mushroom had several name changes throughout the years before landing on the name *Psilocybe cubensis* given in 1949 by German mycologist Rolf Singer. *Psilocybe* is derived from the ancient Greek word meaning "bare head" and *cubensis* meaning "from Cuba," which is where Franklin Sumner Earle described the mushrooms in 1906.

Although they weren't given a Latin name until 1906, these mushrooms have been used by cultures around the world for

Mushrooms in Evolution

Magic Mushrooms are thought to have impacted human growth from apes too. There's even a theory by Terence and Dennis McKenna, called the Stoned Ape Theory, which states that human evolution was sparked by our early ancestors' ritual eating of Magic Mushrooms.

thousands of years. There are cave paintings of *Psilocybe* mushrooms as far back as 4713 B.C.E. and people from all parts of the globe, including in ancient Mexico, used them in ritual practices. The Aztecs and Mayans used these mushrooms and called them *Teonanácatl*, meaning "flesh of the gods" in the Nahuatl language.

Magic Mushrooms took the United States by storm when Gordon and Valeria Wasson traveled to Huautla de Jiménez in Oaxaca, Mexico, and had a Magic Mushroom ceremony with a *curandera* (Spanish for a native healer or shaman) named Maria Sabina. The Wassons published their experiences in *Life* magazine in 1957, introducing the general public in the United States to Magic Mushrooms. Soon after, psychologists Timothy Leary and Richard Alpert created the Harvard Psilocybin Project at Harvard University doing various studies on these mushrooms.

Unfortunately, Psilocybin was banned in 1968 and became a Schedule 1 drug in 1971 in the United States. It continues to be legal in various countries like Brazil, Jamaica, the Bahamas, Nepal, and Samoa.

Magic Mushrooms are actually one of the easiest mushrooms to grow; the average person could easily grow them at home. There is no toxicity in the mushroom, and every year more scientific studies show that shrooms are effective for treating addiction, depression, and anxiety, and for providing end-of-life care. One practice that is gaining popularity is micro-dosing, which is taking 0.1–0.4 grams of Psilocybin at a time. Only a tiny amount of the effects is felt, while getting the benefits of the psilocybin throughout the day.

Maitake

Grifola frondosa

Maitake is also known as Hen of the Woods, Sheep's Head, or Ram's Head. In Japan, it is called "Dancing Mushroom" because when you find it, you would do a little dance of happiness. Maitake was so revered in Japanese culture that the areas in which it grew were called "treasure islands" and the mushrooms were worth their weight in silver. Maitake is pretty popular in the northeastern portion of the United States, where it grows in abundance. Some of these mushrooms grow as large as 100 pounds.

Hen of the Woods grows from a small sclerotium, or mycelial ball, underground in association with tree roots. This mushroom also comes back annually. Since this is such a prized mushroom, people guard their Maitake spots meticulously. Additionally, although there are no deadly look-alikes for Maitake, some people confuse the Black-Staining Polypore with Maitake. The Polypore is still an edible species but is less desirable.

Maitake is most known for its flavor and use in cooking. It's been consumed in China and Japan for centuries. This is one of the most prized culinary mushrooms and it can really enhance a dish. These mushrooms have a strong earthy musk, as well as a peppery taste.

Apart from being found in the wild, these mushrooms can be commercially cultivated. Although a tricky mushroom to grow, the process has been figured out and replicated on a mass scale. Like all functional mushrooms, Maitake are packed with compounds that help support the body's immune system. Here's a list of the wide variety of things that this mushroom's different compounds can do:

+ Hen of the Woods has a popular compound that supports vital protein health.

+ This mushroom has two other compounds that help support cell health and signaling.

+ Maitake also has high levels of a compound that supports the nervous system, neuronal health, and healthy aging.

+ Yet another compound, which supports healthy energy and cell function, is found in greater concentrations in the mushrooms compared to the mycelium.

The most common reasons why people take Maitake for functional benefits are for providing metabolic support, supporting healthy cell turnover, supporting the body in maintaining homeostasis and fighting against occasional stress, as well as supporting the immune system. In addition to its culinary uses, it can be consumed in tinctures, powders, and capsules.

Maitake Recipe Ideas

Maitake, like many gourmet mushrooms, can be cooked in a wide array of delicious ways. Many people use this mushroom as a substitute for meat—a popular dish is Maitake pulled "pork." This mushroom is often used in warm salads/bowls, soups, as well as toppings on gourmet pizzas. Also, simple does not mean worse: Grilling or sautéing these mushrooms are delicious options.

MAITAKE

MATSUTAKE

Matsutake

Tricholoma matsutake

Matsutake, also known as the Buddha Mushroom, is one of the most sought-after edible mushrooms in the world. It fetches a jaw-dropping price of up to $1,000 per kilogram (2.2 pounds). The name Matsutake comes from the Japanese words *matsu*, meaning "pine tree," and *take*, meaning "mushroom." This name is due to the fact that this mushroom forms a mycorrhizal connection with various pine trees. However, the Matsutake can also form relationships with coniferous trees like Douglas fir, noble fir, Shasta red fir, and hardwoods like tan oak, madrone, rhododendron, salal, and manzanita trees.

The North American varieties of Matsutake are *Tricholoma magnivelare* and *Tricholoma murrillianum*, which are very similar in look and taste to the original Matsutake. These North American varieties are routinely used instead of the proper "true" Matsutake (*Tricholoma matsutake*).

Currently, Matsutakes are even more in demand, due to a pest problem in Japan. *Bursaphelenchus xylophilus*, a pine-killing nematode, is devastating the natural pine tree population in Japan that the Matsutake rely on for their symbiotic relationships. From this decreased supply comes a higher demand, and with a higher demand comes a greater dependency on international import. Right now, most of the supply of Matsutake in Japan is coming from North America, China, Korea, and Europe. Because these are not "true" Matsutakes, the imported mushrooms cost much less. The price is dependent on quality, availability of local mushrooms, and origin. Usually around one thousand tons of Matsutake are harvested in Japan each year, but the nematode is throwing a huge wrench in the Japanese mushroom economy. There is actually a whole book written about it called *The Mushroom at the End of the World* by Anna Lowenhaupt Tsing. Tsing writes about the effects of capitalism through the lens of the trade of this mushroom.

Surprisingly enough, this mushroom is revered more for its smell than for its taste. The smell of the Matsutake has stumped mycologists for decades. Many try to describe it and fail. The most accurate description has been…it smells like Matsutake. The smell is unique; it has been variously explained as smelling like dirty socks, cinnamon, woodsy, spicy, earthy, and like a "multidimensional truffle." The taste is much more easily defined. As expected, this mushroom has a pine-forward taste, with an earthy spiciness to it.

Cooking Tips for Matsutake

Because of its high price, gentleness is recommended when cooking with Matsutake; you should serve them either alone or at the center of the dish to enjoy them as much as possible. In Japan the two most popular ways of cooking Matsutake are in a rice dish called Matsutake gohan and in a flavorful broth called sukiyaki.

Meshima

Phellinus linteus

Meshima means "women's island" in Japanese. This is an apt description of this mushroom, as it's used in traditional Japanese, Chinese, and Korean herbalism for supporting overall women's health. Meshima is known as *Song Gen* in Chinese and *Sanghwang* in Korean. It is also commonly called the Black Hoof Mushroom.

This mushroom can be found growing on a variety of places on Mulberry trees: on the trunk and also high on branches. If luck is on your side, you may be able to find the mushroom on a fallen log. On the trees, it emerges from the bark and sits on the side of the tree like a shelf. The actual mushroom grows in these layers that have emerged from the tree. The layers and mushroom then fuse together into a horse hoof shape. The bottom of the mushroom has a porous surface from which the spores disperse.

All functional mushrooms are considered adaptogens, meaning they help support the body's natural ability to deal with occasional stress while being naturally nontoxic. All functional mushrooms also support the immune system because they are packed with a type of compound found specifically in fungi. This compound is well studied and found to be supportive to a healthy functioning immune system.

Because it's a hard conk, a shelf-like mushroom growing from a tree, it would be impossible to cook Meshima as an edible (never mind delicious) meal. Meshima is traditionally prepared in a tea as it's the easiest way to consume this mushroom. The best way to get the largest number of beneficial compounds from this mushroom is to make a tincture. When you make a tincture, you extract the mushroom's helpful compounds with alcohol. The tincture can then be combined with a tea that contains complementary compounds. As with all functional mushrooms, it's best to use the fruiting body as opposed to the mycelium. The mushroom's body has the highest concentration of functional compounds that are supportive for human health. Meshima, much like Reishi Mushrooms, has a very bitter taste. These bitter compounds are a sign of potency; so, the more bitter, the better.

Take Functional Mushrooms As a Daily Routine

Functional mushrooms are best taken at the same time every day, as part of a routine. If you make smoothies every morning, it's easy to add some powdered Meshima to your daily breakfast. Or, add them to your afternoon latte. The most important part is being consistent; that way you are more likely to feel the effects!

MESHIMA

MILK CAPS

Milk Caps

Lactarius spp. and *Lactifluus* spp.

Milk Caps are generally species in the *Lactarius* and *Lactifluus* genera that produce a milky substance when bruised or cut. They are cap-and-stem mushrooms with a brittle texture typical for the family *Russulaceae*. These mushrooms grow all around the world and form ectomycorrhizal symbiotic relationships with a range of trees. Examples of Milk Caps include the following, minus the Indigo Milky Cap (*Lactarius indigo*), which has its own entry in this book.

✦ *Multifurca furcata*: Although most Milk Caps are in the genera *Lactarius* and *Lactifluus*, there is one exception: *Multifurca furcata*, which is the only species in the genus *Multifurca* that produces the milky substance that causes it to be considered a Milk Cap. The *Multifurca furcata* was originally grouped under the genus *Lactarius*, though when viewed under the microscope, it's clear that it belongs in this different, rare genus.

✦ *Lactarius deliciosus*: Also called Saffron Milk Cap or Red Pine Mushroom, this mushroom is one of the best-known Milk Caps. It grows under conifers mainly in Europe and India. The Saffron Milk Cap is a highly sought-after edible mushroom; it's usually fried or pickled. Some consider it mild or bitter, but others love it.

+ *Lactarius deterrimus:* Also called the False Saffron Milk Cap or Orange Milk Cap, this mushroom forms a relationship mostly with Norway spruce trees. It is considered a delicious edible mushroom. Just like with *Lactarius deliciosus*, if you eat a lot of these mushrooms your urine turns red because of the azulene compounds in them.

+ *Lactarius quietus:* Also called Oak Milk Cap, Oakbug Milk Cap, or Southern Milk Cap, this one is an inedible brownish mushroom that smells a bit like oil. This mushroom is exclusive to Europe, and lives mostly under oaks.

+ *Lactarius torminosus:* Also called the Wooly Milk Cap or the Bearded Milk Cap, this mushroom is controversial because, if eaten raw, it can blister the tongue and cause gastroenteritis. However, if parboiled to leach the toxins out, it can be enjoyed in food or in coffee as it's consumed in Europe and Russia.

+ *Lactarius turpis:* Also called the Ugly Milk Cap, this mushroom is not the most attractive mushroom to look at. It's even worse to eat.

+ *Lactifluus volemus:* Also called the Weeping Milk Cap or the Voluminous-Latex Milky, this mushroom has a distinctive fishy smell. The smell disappears when cooked and this is surprisingly a delicious edible mushroom. It also has various compounds that are being researched for their functional benefits and for the purpose of making rubber.

Fairy Rings

Mushrooms have a lot of connections to mythology and storytelling. One of the more famous instances of mushroom fantasy is based in reality. Fairy rings are when mushrooms grow naturally in a circle. They are also called fairy circles, elf circles, or pixie rings. These rings are oftentimes formed by field mushrooms, or *Agaricus campestris*.

Morel Mushroom

Morchella esculenta

AT A GLANCE ⊙

GEOGRAPHIC LOCATION ✦
North, Central, and South America;
Europe; and Asia.

GROWING LOCATION ✦ On the
ground in association with pine,
elm, tulip, sassafras, beech, ash,
sycamore, and hickory trees.

CHARACTERISTICS ✦ This mushroom
has a tannish-white stem leading
toward a pitted, brain-like, tan to
brown to black cap-like structure.

PRIMARY USE ✦ Edible.

Morel Mushrooms are one of the most sought-after wild mushrooms in the world. Though these mushrooms grow on several continents, people guard their Morel spots with their lives. These mushrooms are typically hard to find, especially for beginners, so they are thought of as rare and elusive, making the feeling of finding them even more rewarding. Their globule and interesting brain-like shape and earth-tone colors provide them perfect natural camouflage. Morels blend into the leaf litter of the trees they grow around and almost disappear into the background (even if you are looking right at them). They are one of the earliest mushrooms to appear in the spring and only pop up for a short window as the ground thaws and warms up. This mushroom was originally called *Phallus esculentus* in 1753 by Swedish botanist Carl Linnaeus, but its name was changed not too long after by Swedish mycologist Elias Magnus Fries in 1801 to its current name, *Morchella esculenta*.

Hunting and eating Morels can be hazardous if you are not careful. A common foraging spot for Morels is in old apple orchards, but most apple farmers from the 1890s to the 1960s used lead arsenate pesticides. Since mushrooms can hyperaccumulate heavy metals, this makes harvesting and eating too many Morel Mushrooms from old apple orchards a concern. Lead and arsenic are both deadly to humans. The other concern when cooking Morels is that raw Morels contain hydrazine, a gastrointestinal irritant. You must first parboil or blanch the mushrooms before consuming them and make sure to have proper airflow when cooking. Otherwise, these mushrooms will make you sick.

Since this is such a delicious mushroom and is enjoyed by people all around the world, many attempts have been made to try to artificially cultivate them on a mass scale. The first successful attempts were made in 1901. Now, the production of Morel Mushrooms has completely taken off in China, where thousands of dried tons of these mushrooms are produced every year in outdoor hoop houses in the countryside. Two other companies in the US and the Netherlands have also developed techniques to grow Morels indoors in trays and beds to make this delectable mushroom more affordable to the masses.

How to Cook Morel Mushrooms

There are a lot of delicious ways to cook Morels. Many people fry or sauté the mushrooms, serving them with sauces. They are also very popular in a variety of pasta dishes, or they would be delicious in risotto. Morels complement meats, such as pork chops or burgers. They are a very adaptable mushroom, but do not eat them raw.

MOREL MUSHROOM

OLD MAN *of the* WOODS

Old Man of the Woods

Strobilomyces strobilaceus

Old Man of the Woods is an edible mushroom from Europe, Asia, and North America that grows in deciduous and coniferous forests. It's mostly known for its spiky scaly cap, which makes it stand out from other mushrooms in the forest. Its cap often gets misidentified as a pinecone on the ground, which is how this mushroom got its scientific name: *Strobilomyces* comes from the Greek word *strobilos*, which translates to "pinecone."

Old Man of the Woods was first named *Boletus strobilaceus* in 1770 by Italian naturalist Giovanni Antonio Scopoli. Its name was later changed to its current name, *Strobilomyces strobilaceus*, by British mycologist Miles Joseph Berkeley in 1851. Berkeley was known for creating the *Strobilomyces* genus in 1851.

Besides its cap, Old Man of the Woods Mushrooms have other unusual physical characteristics. Under the cap, this mushroom has geometric tubes that spores come out of, as opposed to the gills found on many common mushrooms. If you press your finger against the pale gray pore surface, the mushroom will bruise black.

Color-Changing Mushroom

Many mushrooms change color as they age, but most do not change color as quickly as the Old Man of the Woods. When you slice this mushroom with a knife, the flesh will turn pinkish to orangish red in about five minutes, and then it will slowly turn a blackish color after about thirty minutes. This is a neat trick to show new mushroom enthusiasts.

This mushroom species has had confusing errors with DNA sequencing and naming over time. These scientific blunders still have not been fully worked out. In Europe, there had been three separate species that were named *Strobilomyces strobilaceus*, *Strobilomyces strobiliformis*, and *Strobilomyces floccopus*, but after DNA analysis they were considered too closely related to have separate names. Therefore, they were all lumped together under the *Strobilomyces strobilaceus* name. In North America, scientists are still working out the Latin names of the *Strobilomyces* species. So, for now, North American mushrooms in this genus include *Strobilomyces floccopus*, *Strobilomyces strobilaceus*, *Strobilomyces confusus*, and *Strobilomyces dryophilus*. With more DNA analysis, these mushrooms might be lumped under one umbrella Latin name or continue to be broken up, but more research needs to be done.

Although the Old Man of the Woods is edible, it's not known having a particularly delicious flavor, and it is not a coveted mushroom. Some chefs suggest sautéing this mushroom with butter or oil like any other culinary mushroom. However, it may be better to get more creative with your recipes to bring out the best in the flavors of this mushroom. It is described as having a muted earthy taste, or a taste similar to common store-bought mushrooms. The good thing about this mushroom, especially if you're in North America, is that it's unique and it's hard to misidentify it.

Oyster Mushroom

 Pleurotus ostreatus

If there is one group of mushrooms that could help the world become more sustainable, fight hunger, aid in nutrient cycling, degrade oil spills, and more, the Oyster Mushrooms would be the candidate. Oyster Mushrooms are among the most well-known mushrooms in the world and are some of the easiest, if not the easiest, mushrooms to cultivate. Oyster Mushrooms have a great flavor, and they are sought after in farmers' markets as a great gourmet mushroom.

Oyster Mushrooms were first described in 1775 by Dutch-born scientist Nikolaus Joseph Freiherr von Jacquin. However, these mushrooms were officially classified as *Pleurotus* in 1871 by German mycologist Paul Kummer.

Since this is a saprophytic mushroom (a fungus that decomposes wood), you can see this mushroom growing on trees that are dead or dying. The Oyster Mushroom is a white rot fungus, meaning it decays an important polymer, lignin, primarily in trees. Lignin helps with the structural stability of trees. This rotting capability makes Oyster Mushrooms a good candidate for mycoremediation; they also have been shown to be active against various biological contaminants. In terms of chemicals, the

Oyster Mushrooms are used to treat wastewater from the pulp and paper industry, as well as water contaminated with pesticides or PCPs. Alex Dorr, the author, used enzymes from *Pleurotus ostreatus* to remediate polycyclic aromatic hydrocarbons (PAHs) from cigarette butts and found 100 percent reduction in many pollutants in just a couple of weeks. The author also did research on the hyperaccumulation of heavy metals from cigarette butts by Oyster Mushrooms and found the mushrooms hyperaccumulated lead, nickel, zinc, copper, aluminum, iron, and manganese. Dorr found that the levels of copper and iron in the mushrooms were higher than amounts considered safe to consume. Therefore, it's important to be careful of finding these mushrooms in the wild—they may have accumulated too many heavy metals for human consumption.

Oyster Mushrooms are packed with many elements that are beneficial for the human body, including amino acids, protein, minerals, fiber, and carbohydrates. Oyster Mushrooms are usually grown commercially on straw or sawdust, but they can be grown on pretty much anything, including:

+ coffee grounds
+ shirts
+ shoes
+ and more

The most common species of Oyster Mushroom is *Pleurotus ostreatus*; other species include *Pleurotus ostreatus* var. *columbinus* (also known as the Blue Oyster), *Pleurotus citrinopileatus* (Golden Oyster or Yellow Oyster), *Pleurotus djamor* (Pink Oyster), *Pleurotus pulmonarius* (Phoenix Oyster), *Pleurotus eryngii* (King Oyster), and lastly the *Hypsizygus ulmarius* (the Elm Oyster; although not a *Pleurotus* species it is still considered an Oyster Mushroom).

Oyster Mushroom Recipes

Oyster Mushrooms are known for their delicious, slightly seafood-like taste, with notes of earthiness and anise. This makes them a great contender for any seafood replacement. Try Oyster Mushrooms in a vegetarian (or meat-inclusive) paella, deep-fry them and serve them with sauce, or throw them in a rice bowl with some greens. These mushrooms are easy to build a variety of delicious dishes around.

OYSTER MUSHROOM

PADDY STRAW MUSHROOM

Paddy Straw Mushroom

Volvariella volvacea

*P*addy Straw Mushrooms, also known as Straw Mushrooms, are a primarily Asian variety of mushrooms that grow off straw. They are one of the most commonly consumed mushrooms in the world. Although they are native to Asia, they were recently brought to North America. In North America, they are the most popular east of the Great Plains in the United States.

In Asia, Straw Mushrooms are usually picked in the egg-like stage before they have erupted and are labeled "unpeeled." In North America, this unpeeled stage can be problematic, as Straw Mushrooms have a deadly look-alike: the Death Cap Mushroom. The Death Caps also resemble an egg in their early stages. This has been an issue for immigrants from Asia who move to the United States or Australia, where Death Caps are abundant and easy to misidentify. The easiest way to tell these two mushrooms apart is to take a spore print: The Paddy Straw Mushrooms have pink spores, while the Death Caps have white spores. In any case, it's wise to consult experts before ingesting. Your best bet is to just head to your local specialty food store or farmers' market to see if these mushrooms are available.

The Paddy Straw Mushrooms are saprophytic mushrooms that grow on decaying organic matter like wood chips, in gardens, on compost piles, or in straw. Straw remains their preferred artificially cultivated habitat. They prefer temperatures of about 90°F, with lots of humidity and moisture. Hot tropical areas of Southeast Asia are the best for cultivating this mushroom. It was originally first grown in China around 1822 and then brought to Southeast Asia around 1932.

The mushrooms grow very quickly, between 5–7 days from spawning, and after they make the "egg," they will reach full maturity after 4–5 more days. They are grown in beds of tied straw bales and can yield 9–11 pounds of fresh mushrooms per 65 pounds of straw.

Straw Mushrooms Are Full of Health Benefits

Paddy Straw Mushrooms are incredibly good for you, being around 14 percent protein fresh and 40 percent protein dry. These mushrooms are also packed with nutrients and minerals like selenium, iron, copper, folate, phosphorus, pantothenic acid, dietary fiber, zinc, and other B vitamins.

The flavor is really sweet and makes a great addition to dishes that are bitter, sour, or salty. These mushrooms are traditionally thrown into hot-and-sour soup or egg drop soup; they can also be added to curries, stir-fried, sautéed, or put into lettuce wraps. They also keep their flavor when canned.

Parasol Mushroom

Macrolepiota procera

The Parasol Mushroom is a famous edible mushroom that grows all around the world. It originally was called *Agaricus procerus* in 1772 by Italian naturalist Giovanni Antonio Scopoli. In 1948 the name was changed to *Macrolepiota procera* by German mycologist Rolf Singer. *Procera* means "tall" in Latin. This is an apt description of these mushrooms, as they typically have a very long stem.

Parasol Mushrooms are commonly found in grassy pastures or sometimes in wooded areas. They can be easily identified by the scaly exterior and notable nipple-like bump on the top center of the cap called an umbo. The scaly exterior of the Parasol Mushroom is sometimes described as snake-like, hence its other European names: Snake's Hat or Snake's Sponge. The common name Parasol comes from the Italian *para il sole*, meaning "keep the sun off," which refers to its affinity to shady locations and the cap when it's fully mature, when it looks like an open umbrella.

In the wild, this is a saprophytic mushroom decomposing organic matter, making it easy to artificially cultivate. You can buy mycelium bags online and inoculate them into your garden using a fresh wood chip substrate. The mushrooms take quite a while to fruit, taking as much as two years before you see any results. Research is currently being done to improve these times and increase mushroom crop yields. China has the most advanced cultivation techniques. However, the Parasol Mushroom is expanding in different uses, as well as gaining appreciation globally, so it will potentially be grown in higher amounts elsewhere in years to come.

Expert foragers can easily identify Parasol Mushrooms, but beginners might confuse them with:

* *Amanita* species such as *A. phalloides* and *A. pantherina*
* *Chlorophyllum* species such as *C. rhacodes* and *C. molybdites*
* Other *Lepiota* species such as *L. aspera*, *L. brunneoincarnata*, *L. helveola*, and *L. pseudolilacea*

As with any mushroom, it's important to have a positive identification before ingesting them. Parasol Mushrooms are mostly picked or cultivated for consuming, as they are delicious edible mushrooms. One popular preparation is to take the caps once they have fully matured and expanded, cut them off, put them in an egg wash, bread them, and bake them to make some delicious breaded mushroom patties. However, the Parasol Mushroom is also just starting to be used for its functional benefits for health and wellness, specifically for its antioxidant support and its supporting qualities on the immune system. More research needs to be done on its functional benefits in addition to how to cultivate this mushroom best artificially.

Parasol Mushroom Folklore

Parasol Mushrooms have an adorable background story in European folklore. In Scotland, it's said that fairies would use them as tables or chairs, while Welsh fairies supposedly used these mushrooms as umbrellas. These mushrooms grace European artwork, as well as a 1995 Azerbaijanian stamp.

PARASOL MUSHROOM

PARROT WAXCAP

Parrot Waxcap

Gliophorus psittacinus

The Parrot Waxcap, also called the Parrot Mushroom, was first named *Agaricus psittacinus* in 1762 by German mycologist Jacob Christian Schäffer. Its Latin name was then changed to *Hygrocybe psittacinus* in 1871 by another German mycologist, Paul Kummer. The mushroom has since changed names again, to its current Latin name given to it by American botanist D. Jean Lodge and collaborators in 2013. The researchers in 2013 advised that further DNA sequencing needs to be done, as they assume *Gliophorus psittacinus* refers to an umbrella of many species—all of these species need to be further broken apart.

Since this mushroom is found throughout Europe, the Americas, Africa, and Japan, and it has a large diversity of characteristics, it's likely that the mushrooms falling under the umbrella of the Parrot Waxcap are separate, but closely related, species. Until it's proven that these Waxcaps are different species, these mushrooms are all given the Latin name *Gliophorus psittacinus*.

An example of the process of eliminating a species from the group of Parrot Waxcaps would be a mushroom called *Gliophorus perplexus*. This mushroom was formerly called *Gliophorus psittacina* var. *perplexa* and before that *Hygrocybe perplexa*. The last name in the list indicates that this species used to be considered a Parrot Waxcap Mushroom variety, but further DNA sequencing kicked it out of that group and gave it its own Latin name. A common joke in the mycology community is: How many mycologists does it take to change a lightbulb? Answer: Fifty. One to change the lightbulb and forty-nine to stand around and argue about the correct Latin name. A fair warning: *Gliophorus psittacinus* might not be the final Latin name to be given to this mushroom, so stay tuned.

The Parrot Waxcap Mushroom is usually found in grassy or mossy areas near wooded roadsides, in lawns, or in graveyards. It's even considered an indicator of grassland quality, as this mushroom commonly appears after years of low-nutrient management.

Gliophorus comes from the Greek *glia*, meaning "glue," and *phoros*, meaning "bearing." The name refers to the thick, slimy liquid that coats the mushroom, which is why this mushroom is referred to as a waxcap. *Psittacinus* comes from the Greek *psittakos*, meaning "parrot," referring to the colors of the mushroom that can be similar to those of the bird's green, yellow, and pink feathers.

Since the Parrot Waxcap is extremely small and covered in this slime, eating this mushroom is not enjoyable and it won't fill you up. This mushroom is also mild in taste and smell. Some reports indicate that people get a slightly upset stomach when eating more than twenty mushrooms, so eating them is not advised. It does not seem worth the trouble.

 A Genus of Many Colors

Though it's likely that this genus of mushroom is in fact made up of a variety of slightly related fungi, the beautiful changing colors should still be admired. One species starts bright green, then changes to a duller yellow, then eventually to a tan, straw-colored mushroom. Another species turns from blue to red over time. There are a large variety of colors throughout this genus, just as there are parrots of many colors.

Pinecone Tooth Mushroom

Auriscalpium vulgare

AT A GLANCE

GEOGRAPHIC LOCATION ✦ Europe, Central and North America, and temperate Asia.

GROWING LOCATION ✦ Conifer litter and conifer cones.

CHARACTERISTICS ✦ Tiny mushrooms that have a bristly fuzzy surface and teeth beneath the cap.

PRIMARY USE ✦ Potentially edible.

The Pinecone Tooth Mushroom, also called the Cone Tooth or Ear-Pick Fungus, is a tiny furry mushroom that likes to grow on pinecones. It typically favors Scots pine, spruce, and Douglas fir pinecones, and, in northeast India, it prefers burned Khasi pine pinecones. A study done in China showed that this mushroom loves to grow at altitudes of 8,500–9,800 feet in mixed forests.

Pinecone Tooth Mushrooms are almost impossible to misidentify: They are the only furry and toothy small mushrooms that like to grow on conifer cones in Europe, Central and North America, and temperate Asia. Although there are a few species, including *Strobilurus trullisatus*, *Baeospora myosura*, and *Mycena purpureofusca*, that grow on conifer cones in the wild, *Auriscalpium vulgare* is the only pinecone mushroom that has teeth as opposed to gills. When young they are white to purplish pink and as they get older, they shift to a brown color.

The species was first described by Swedish botanist Carl Linnaeus in 1753 as *Hydnum auriscalpium*. The Pinecone Tooth Mushroom's name later changed to its current *Auriscalpium vulgare* when

British botanist Samuel Frederick Gray renamed it in 1821. *Vulgare* comes from the Latin word for "common," whereas *Auriscalpium* comes from the Latin for "ear pick," referring to the shape of the mushroom. The mushroom looks like a small scoop-shaped instrument that was commonly used to clean ears at the time.

In the wild this fungus is fairly picky, needing high levels of humidity to produce mushrooms. It also requires the perfect amount of light. In sterile cultures, Pinecone Tooth Mushrooms can be grown on agar and fruit with optimal conditions. Mushrooms being able to fruit on agar is rare. However, this species is able to fruit just six weeks after they are first inoculated onto the agar plate. They then take about sixty days to fully mature once they start to form.

Although it's considered inedible by most mushroom lovers because of its bristly texture and small size, Pinecone Tooth Mushrooms historically were eaten in Italy and France. That said, it's probably not worth eating due to its mild taste and smell.

Prehistoric Mushrooms the Size of Trees

It seems hard to believe, but before the world was covered in forests, large fungal structures called prototaxites loomed over the planet's surface. These fungal structures were up to 24 feet tall. Scientists have uncovered fossils of these organisms and found that they were, in fact, mostly fungal.

Many foragers have a hard time finding this mushroom as it tends to grow on partially buried conifer cones. The Pinecone Tooth Mushroom is a tiny mushroom (only 0.29–2.76 inches tall); only 1–5 mushrooms will grow on a cone. Some people also find them on spruce needles above squirrel dens.

PINECONE TOOTH MUSHROOM

POISON FIRE CORAL

Poison Fire Coral

 *Podostroma cornu-damae*

*P*oison Fire Coral, *Podostroma cornu-damae*, is one of the deadliest mushrooms in the world. The Japanese name, *Kaentake*, means "flame fungus," from its fiery-red color. This mushroom was originally called *Hypocrea cornu-damae* in 1895 by French mycologist Narcisse Théophile Patouillard. The mushroom was then transferred into the *Podocrea* genus by Italian mycologist Pier Andrea Saccardo in 1905. It was then finally given its current name, *Podostroma cornu-damae*, in 1994 by Japanese mycologists Masana Izawa and Tsuguo Hongo.

Originally native to Korea and Japan, the Poison Fire Coral has started to pop up in Indonesia, Papua New Guinea, and Australia, scaring locals due to the deadly repercussions of people picking and eating this mushroom. In Australia, this mushroom was found in 2019, where locals added it to the already long list of potentially fatal organisms on the continent.

When the mushroom is young it can be confused for *Ganoderma lucidum*, *Cordyceps* species, or *Clavulinopsis miyabeana*, which has also been known to be fatal if ingested. Some of the mushrooms that the Poison Fire Coral has been mistaken for are used in traditional Chinese herbalism, which is obviously very problematic.

The Poison Fire Coral Mushroom might be the only mushroom in the world to cause irritation from just touching it. Some claim that people can develop skin irritations and rashes after touching it; others believe that's just fearmongering. Beyond being an irritant when touched, this mushroom can be deadly when eaten. The main method of action for the poisonings come from several different toxins, all of which if injected into mice at only a 500-microgram dose will kill them within twenty-four hours. In humans the usual symptoms of ingesting this mushroom include stomach pains, peeling skin on the face, hair loss, vomiting, diarrhea, fever, numbness, and a shrinking cerebellum, which results in changes in perception, speech impairments, and problems with voluntary movements. If not treated, poisoning can lead to multiple organ failure, blood disorders, brain damage, and death.

In order to recover from eating one of these mushrooms, you must go to the ER. The treatment is likened to someone having chemotherapy and bone marrow suppression. Treatments include blood transfusions. A patient will also need to be kept in isolation during treatment.

Mushrooms As Protectors

Many mushrooms actually help out the human body. Some mushrooms support the immune system. Other mushrooms are believed to support antioxidant activity and our bodies' resilience against occasional stress and fatigue. Some others help with energy levels, as well as other functional benefits.

Porcini

Boletus edulis

*P*orcini is also called Penny Bun or King Bolete. Throughout the world, Porcini has other fascinating names and meanings:

✦ It's called *Steinpilz* in Germany, meaning "stone mushroom."
✦ In Austria it's known as *Herrenpilz*, which means "noble mushroom."
✦ In Mexico the name is *Panza* or *Rodellon*, which mean "belly" and "small round boulder" respectively.
✦ In Dutch it's called *Eekhoorntjesbrood*, meaning "squirrel's bread."
✦ In Russian it's called *Belyy Grib*, meaning "white mushroom."
✦ It's called *Cep* in Catalan.
✦ French people call this mushroom *Cèpe de Bordeaux*.
✦ In Italy, it's called *Cappatello Buono*.
✦ It's *Herkkutatti* in Finish, meaning "delicious mushroom."
✦ Both names, *Kaljohanssvamp* in Swedish and *Karl Johan Svamp* in Danish, honor King Charles XIV John who popularized the mushroom.

Porcini is one of the most prized edible mushrooms in the world and grows all over Europe, North America, Africa, Asia, Australia, New Zealand, and Brazil, making symbiotic mycorrhizal connections with pine, spruce, hemlock, fir, chestnut, chinquapin, beech, *Keteleeria* spp., and oak trees. It was first described in 1782 by French botanist Jean Baptiste François Pierre Bulliard. *Boletus* is derived from Greek, and it means "terrestrial fungus." *Edulis* is the Latin word for "edible."

Being one of the most prized edible mushroom species in the world, it's also luckily easy to identify. There are only a couple look-alike species, but a trained eye can easily differentiate between them. The first look-alike is the poisonous Devil's Bolete, which although it has the same bulbous shape, can be easily told apart based on the red stem and blue bruising of the Devil's Bolete. The other look-alike is *Tylopilus felleus*, which is incredibly bitter. The easiest way to tell the difference is to nibble on a tiny amount and quickly spit it out. If the mushroom is bitter, you know it's not Porcini. *Gyroporus castaneus* is another look-alike which fortunately is edible, except for a select poisonous strain off the coast of Portugal. Use caution if you live there or plan to forage there. The main difference is that the spores of *Gyroporus castaneus* are a light straw color, whereas the true Porcini mushrooms have dark, olive-brown spores.

How to Select Good Porcini

At the farmers' market, use these tips to identify good Porcini mushrooms: Avoid mushrooms with black spots or holes. The caps of the mushrooms should be large, undamaged, and brown in color. Bruises on mushrooms indicate they will go bad quickly, if they haven't already.

Once you have identified *Boletus edulis*, make sure to pick it young, as insect larvae also find it delicious. The mushroom has a nutty taste, with some earthiness. It tastes phenomenal in pasta dishes, sautéed in butter, in soups, risotto, and more. It's sold fresh, dried, pickled, frozen or incorporated in butter. Up to 100,000 tons are consumed annually around the world, for good reason. Currently many farmers and scientists are trying to figure out how to artificially cultivate this mushroom, but no one has been successful yet.

PORCINI

PORIA COCOS

Poria Cocos

Wolfiporia extensa

Not technically a mushroom, Poria Cocos is considered a sclerotium or a hardened fungal mass that lives underground. This fungus resembles a large yam underground and traditionally they are either sold at the market in white strips or in white cubes. Other names for this fungus include Hoelen, Poria, Chinese Tuckahoe, China Root, Matsuhodo, or the most common name in China: Fu Ling. This fungus has a long history in traditional Chinese herbalism for its use in keeping the human body healthy.

Poria Cocos is distributed in Asia as well as in Mexico and the United States. Saprophytic in nature, these fungi feed on dead or decaying wood and live in relation to specific tree roots such as pine, citrus, magnolia, and eucalyptus. These fungi can also be artificially cultivated on buried logs underground. This is actually a common practice on farms in Asia to supplement the market.

Poria Cocos is studied for its extensive number of specific acids that help support the body's natural inflammatory response after a workout or other normal activities. One of the most useful compounds in this fungus is pachymic acid, which has been well documented for supporting the immune system. Since this is a functional fungus, it's also classified as an adaptogen, having supportive qualities for our bodies' natural ability to deal with occasional stress. In addition to the special acids, Poria Cocos are also known for supporting a healthy immune system. This fungus's uses in traditional Chinese herbalism are extensive. Practitioners use Poria Cocos to help support a healthy mood and promote a better sense of well-being. This fungus is also used for supporting a healthy urinary tract and draining dampness. The outer brown to reddish layer of the sclerotium is mainly used for this as it contains compounds supportive for healthy urinary tract health. As with most functional mushrooms, it's best to use the fruiting body, or in this case the sclerotium instead of the raw mycelium as it's been shown to house more of the beneficial compounds. For example, the sclerotium of Poria Cocos has ninety-one healthy acids, while the mycelium only has nineteen. Also, a good rule of thumb when consuming functional mushrooms or fungi is that they contain both polar and nonpolar compounds, so it's best to extract with alcohol and hot water to pull all the beneficial compounds out, instead of just making a tea.

Not a Native American Tuckahoe

This fungus, though one of its names is Chinese Tuckahoe, is not to be confused with the Tuckahoe that many Native Americans used as a food source. *Tuckahoe* is a generic term for various edible plant tubers, such as *Orontium aquaicum* or *Peltandra virginica*, but is also used for this fungus.

Puffballs

Bovista spp., *Calbovista* spp., *Calvatia* spp.,
Handkea spp., *Lycoperdon* spp., etc.

*P*uffballs is an umbrella term for mushrooms that have a sphere-like shape called a gasterothecium, from the word *gasteroid*, meaning "stomach-like." This sphere shape is also a protective outer structure, called a peridium. The peridium houses the spores and this peridium makes the Puffballs unique. In the case of many species of these Puffball Mushrooms, when raindrops land on the fruiting bodies, they will give off a puff of spores that looks like smoke. Hence their name.

Puffballs are generally split into three sections: Stalked Puffballs, True Puffballs, and False Puffballs. There are only a few Stalked Puffballs, which include *Battarrea phalloides*, *Calostoma cinnabarina*, *Pisolithus tinctorius*, and the genus *Tulostoma*. True Puffballs include the *Bovista* spp., *Calvatia* spp., *Handkea* spp., and *Lycoperdon* spp. Species of False Puffballs include *Endoptychum agaricoides*, *Nivatogastrium nubigenum*, *Podaxis pistillaris*, *Rhizopogon rubescens*, *Truncocolumella citrina*, and potentially other species in their egg-like stage (such as stinkhorns or *Amanita*). Some might argue that species in the

Geastrum or *Scleroderma* genera would be considered Puffballs, but these genera are Earthstar fungi and Earthball fungi, respectively. Earthstar and Earthball fungi are related to the Puffballs but differ slightly.

When foraging for edible Puffballs, a good rule of thumb is to always cut it open. Edible Puffballs should be pure white and firm inside. If it's purplish, black, yellow, green, or any other color, stay away. If you cut it open and you can see an outline of a cap-and-stem mushroom in the spherical shape, this is probably either an *Amanita* or stinkhorn "egg," which could be deadly.

Some common types of Puffballs are:

+ Sculpted Puffball (*Calvatia sculpta*) is a spiky-looking Puffball that is edible when young when the flesh inside is white. The Sierra Miwok Native Americans called this fungus *Potokele* or *Patapsi*.

+ Giant Puffball (*Calvatia gigantea*) is sometimes confused with a giant soccer ball but is an edible mushroom when young. It can be cooked like tofu. It's also used to support the immune system.

+ Brain Puffball (*Calvatia craniiformis*) is an edible brain-like Puffball that is used by Ojibwe peoples and in Chinese and Japanese herbalism as a hemostatic agent to stop bleeding.

+ Common Puffball (*Lycoperdon perlatum*), also referred to as Poor Man's Sweetbread, can be cooked in soups as a substitute for dumplings and are delicious mushrooms. These mushrooms are known as having a mild flavor but a delightful, spongy texture.

Puffball Ink

As with some species of colorful mushrooms, such as the Lobster Mushroom, Puffballs, though muted in color, can actually be used for ink. In Tibet, people would burn these mushrooms and grind down the ash. Then, they would add some liquids and let the mixture sit, pressing it until it became an ink.

PUFFBALLS

RED-BELTED CONK

Red-Belted Conk

Fomitopsis pinicola

The Red-Belted Conk is a saprophytic mushroom, meaning it grows on dead or decaying wood, particularly conifers. It is found in North America and the temperate regions of Europe and Asia. The mushroom itself is very hard and forms a hoof-like conk on the sides of trees or roots of trees. The various colored bands may range from yellow to black to orange to white to red.

Though it's unclear who first described this mushroom as *Boletus pinicola* in 1810, it was then described by Petter Adolf Karsten, Finnish mycologist, in 1881 and classified to the genus *Fomitopsis*.

Ecologically this mushroom acts as what's called a brown-rot, or heartrot, fungus. This means that the Red-Belted Conk decays cellulose in trees and leaves the wood more brittle and more prone to breakage. The tree then becomes unsuitable for pulp production. This process of rotting, however, is important for other animals like woodpeckers, voles, squirrels, bears, and a few other bird species who can utilize this less-dense wood to burrow into the weakened trees, making the trees habitable. Because this mushroom degrades wood in the environment, nutrient cycling in the ecosystem may continue, meaning that more organisms are able to benefit from the nutrients.

Due to increased DNA testing capabilities, the Red-Belted Conk is currently being split up into more specific species. A few examples of these species are *F. mounceae*, *F. ochracea*, and *F. schrenkii*. The DNA testing is done to differentiate the species spread across North America, Europe, and Asia. Latin names for mushrooms change all the time as more DNA testing is completed.

The Red-Belted Conk is known by the Cree of Eastern Canada as *Mech Quah Too*, or "Red Touchwood." It has a medical-usage history. This mushroom was used traditionally in a powdered form to stop bleeding. The mushroom itself is packed with compounds that have shown impressive supportive benefits for human health. Certain compounds within the mushroom have been shown to be supportive for the immune system, as well as be supportive for optimal liver health. In Japan, this mushroom is referred to as *Tsugasaruno-Koshikake* and has been used there traditionally to support the immune system. The tea is thought to be mild, sweet, and only has a small amount of bitter aftertaste to it. This mushroom has also been used in homeopathy, cosmetics, dyes, biomaterials, essential oils, additives to whiskey, and mycoremediation for gold removal.

Mushrooms As Packaging Material

A few major companies worldwide are looking for organic Styrofoam alternatives. No longer wanting to contribute to the wastefulness of current packaging methods, companies are beginning to turn to using mycelium as packaging material. Mycelium, or the "root" of the mushroom, is dense but lightweight, and fully biodegradable.

Reishi

Ganoderma lingzhi

Reishi is an umbrella term for many related mushrooms in the *Ganoderma* genus. These mushrooms are commonly referred to as *Ganoderma lucidum* in literature, but with more DNA analysis this mushroom was renamed. In China, it was renamed to *Ganoderma lingzhi*. In Europe, the species is *Ganoderma lucidum*. And North America has a few other species like *G. tsugae*, *G. oregonense*, and others. Further DNA analysis will split these mushrooms up even more.

The name of the genus, *Ganoderma*, breaks into two parts: *gan* translated as "shiny" and *derm* translated as "skin." These mushrooms have a thin shiny lacquer on the cap that gives off a beautiful and bright shiny red color. The underside is pure white with tons of tiny pores where the spores come out. If multiple mushrooms are growing together, you can see the tops of the caps covered with the brown spores of the other mushrooms. Even spores of the Reishi are harvested and used functionally for health and wellness, as they are packed with beneficial compounds.

Reishi is one of the most well-studied and most cultivated functional mushrooms in the world. There are a few popular methods of growing this mushroom. It can be cultivated indoors in hardwood sawdust bags, or outside in buried logs in hoop houses. Hoop houses are artificially made aboveground tunnels that are covered in shade cloth, keeping the area partially shaded and higher in humidity for the mushrooms to grow. Once these mushrooms dry, they are pretty resistant to rotting and can be used as a sculpture in your house for many years without worry it will mold or rot.

Reishi is also called the Ten Thousand Year Mushroom or the Mushroom of Immortality. It is referred to as the number one mushroom in traditional Chinese herbalism and has been used for thousands of years. Reishi is packed with nutrients that are very supportive for the immune system, but it is most well-known for specific compounds called triterpenes. These compounds help with immune support, seasonal transition support, cardiovascular support, dealing with occasional stress, skin health, and supporting a healthy inflammatory response post workout. Since Reishi is packed with triterpenes, which are nonpolar compounds, these compounds are best extracted with alcohol. The most common method for enjoying Reishi is tea, but this will only pull a fraction of the beneficial compounds, therefore utilizing a double extraction (both alcohol and hot water) is recommended.

Mushroom Art

Bonsai trees are grown and meticulously kept as almost a form of art. Some skilled growers can similarly manipulate the growing of Reishi to make Reishi bonsai sculptures. This is another way to celebrate these species of beautiful mushrooms, which can be put on display.

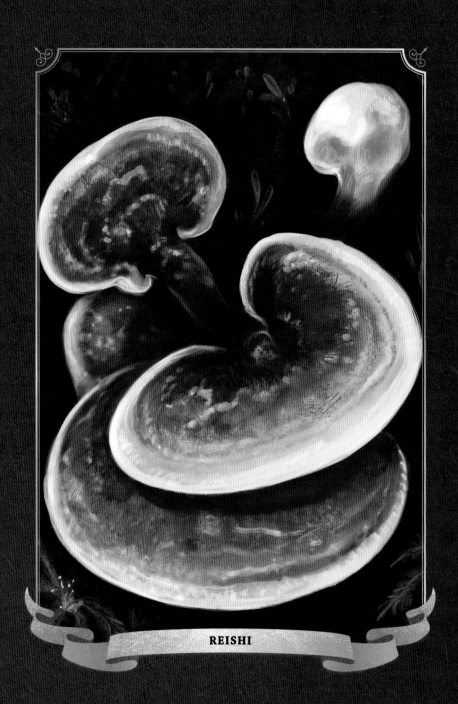

REISHI

SCARLET CUP

Scarlet Cup

Sarcoscypha coccinea

AT A GLANCE

GEOGRAPHIC LOCATION ✦ North and South America, Asia, Africa, Australia, and Europe.

GROWING LOCATION ✦ Found on sticks in damp spots on the forest floor surrounded by leaf litter.

CHARACTERISTICS ✦ A small pinkish-red cup-like mushroom.

PRIMARY USE ✦ Edible, decoration, functional.

Scarlet Cup, also called Scarlet Elf Cup or Scarlet Elf Cap, is a cup mushroom in the family Sacroscyphaceae. The pinkish-red cup or curled saucer-shaped mushrooms grow from sticks and wood debris on the ground, using their enzymes to break down the woody material into food. The red color is caused by five types of carotene pigments; these pigments give the mushroom its name: Scarlet Cup. It's found all over the world in damp underbrush.

Scarlet Elf Cup is one of the only mushrooms known for an audible "puffing" sound, which happens when a cloud of thousands of spores is discharged.

Originally named *Helvella coccinea* in 1772 by Italian naturalist Giovanni Antonio Scopoli, the mushroom was then named *Peziza coccinea* in 1774 by Dutch-born scientist Nikolaus Joseph Freiherr von Jacquin. The Scarlet Cup was then renamed *Peziza dichroa* in 1799 by Danish botanist Theodor Holmskjold, and then finally in 1889 it was given its current name, *Sarcoscypha coccinea*, by Belgian mycologist Jean Baptiste Émil Lambotte. As with most mushrooms, the Latin name of this species is constantly changing as more DNA testing is completed. DNA testing reveals that many mushrooms that were once thrown under one Latin name are actually different mushroom species.

Like many things in mycology, this mushroom is a hot topic of debate. When mycologists talk about the edibility of this mushroom, they come to different conclusions. Although the Scarlet Elf Cap is technically edible, some guides say "inedible" or "not recommended" probably due to its small size. It's described as having an earthy, beet-like taste, and people claim that these mushrooms cook up nicely. It's commonly eaten in France and Switzerland, where people eat it raw on bread with butter, or in fruit salad with kirsch (a sour brandy). Online, you can find many recipes for this mushroom such as:

+ Stuffed Scarlet Cups
+ Omelets
+ Pickled Scarlet Elf Cups
+ and more

The Scarlet Cup was used as a functional mushroom by the Oneida people, one of the founding nations of the Iroquois Confederacy. They would dry it, powder it, and use it to stop bleeding. The Oneida especially used it to stop the navels of newborn children from bleeding after the umbilical cord was cut.

Scarlet Cup Mushroom Look-Alikes

As many mushrooms have look-alikes, so does the Scarlet Elf Cup. Many of the look-alikes are within this mushroom's genus. Though it's less common in nature, the Ruby Elf Cup looks very similar, and it's edible. Another species, the Orange Peel Fungus, looks very similar as well, and is also edible.

In Scarborough, England, they don't eat it or use it functionally, but they put Scarlet Cups in moss arrangements for table decorations. The Scarlet Elf Cap is currently classified as a threatened species in Europe and in Turkey it's critically endangered. So, although it's edible, do not seek it out and use it for cooking in those areas.

Shaggy Mane

Coprinus comatus

*S*haggy Mane, also known as Shaggy Ink Cap, Lawyer's Wig, or Inky Cap, is an interesting edible mushroom that, once picked, is only viable to eat for a couple hours before dissolving into a black ink-like substance.

This mushroom was first called *Agaricus comatus* in 1780 by Danish naturalist Otto Friedrich Müller, and then seventeen years later in 1797 it was given its current Latin name, *Coprinus comatus*, by German mycologist Christiaan Hendrik Persoon. The name *comatus* comes from the Latin *coma* meaning "hair," describing the shaggy hair-like scales on the cap, and *Coprinus* means "living on dung." The Shaggy Ink Cap grows in North America, Europe, Asia, Australia, New Zealand, and Iceland, typically in grassy, disturbed areas typically by footpaths, lawns, or in woodland areas. Australia incorporated this mushroom on a postage stamp in 1981.

The Shaggy Mane Mushroom has a very notable appearance when young, being a scaly or shaggy all-white mushroom that looks like an elongated sphere on a stem. Once they start to age, Shaggy

Mane Mushrooms quickly disintegrate from the edge of the cap, turning into a black ink-like substance.

Since it has such a short shelf life, it's advised to cook this mushroom as soon as possible after you pluck it from the ground. Some chefs have opted to microwaving them into an inky consistency and then freezing them for later use as the liquid component in risotto instead of chicken stock. Since cooking them produces a lot of liquid, they are mostly cooked in soups and broths but can also be sautéed. Some chefs have also gotten creative, incorporating this mushroom into butter.

Various studies show that the Shaggy Mane Mushroom can:

+ help support the immune system
+ support the body's natural inflammatory response post workout
+ keep the body's blood sugar levels in normal ranges
+ provide antioxidant support
+ support liver health

It's advised when picking this mushroom to be careful about foraging near industrial areas or roadways, as it is known to be really good at hyperaccumulating heavy metals in the soil. This is bad news for mushroom pickers but could be good news for potential environmental remediators looking to clean heavy-metal-ridden soil. Shaggy Mane Mushrooms can also be used as a nematicide, as they kill and digest two nematode species: *Panagrellus redivivus* and *Meloidogyne arenaria*.

Why the Shaggy Mane Destroys Itself

Why would a fungus go through the hard work of producing a mushroom, just so that the mushroom then causes itself to melt? It's a different form of evolution. The Shaggy Mane does not rely on other environmental factors to disperse its spores. As it "melts," the mushroom opens wide and releases its spores into its surrounding environment, where it will hopefully reproduce.

SHAGGY MANE

SHIITAKE

Shiitake

Lentinula edodes

The name Shiitake comes from the Japanese *Shii*, the name for the Japanese chinquapin tree, which it commonly grows on, and *take*, meaning "mushroom." Shiitake can also grow on chestnuts, oaks, maples, beech, sweet gum, poplar, hornbeam, ironwood, and mulberry trees.

Shiitake was one of the earliest mushrooms to be artificially cultivated, starting in 1209 during the Song Dynasty in China. He Zhan wrote a 185-word description of the process of cultivating Shiitake Mushrooms. This description was eventually turned into a book in 1796 by Japanese horticulturist Satō Chūryō.

Shiitake Mushrooms are a hot commodity and even have different grades of quality, the highest grade being *donko* in Japanese, or *dōnggū* in Chinese. This highest grade is reached when the mushrooms are introduced to changing temperatures, leaving the cap to crack, which makes beautiful patterns on the top and around the edge. Some chefs say these mushrooms trap flavor in their cracks and, therefore, these mushrooms are of higher quality. The grading system also holds true with outdoor log-grown Shiitake versus indoor bag-grown Shiitake. The outdoor-grown mushrooms have meatier and earthier tasting fruits, while the indoor grown are typically less meaty, flimsier, and have less flavor overall. For those interested in mushroom growing, most will grow Shiitake outdoors in logs that they stack like log cabins, or they will lean them against a fence or tree. These growing methods will produce Shiitakes a few times a season.

Shiitake is a delicious gourmet mushroom that tastes amazing fried in butter or oil to a golden-brown, crispy umami texture and consistency. There are a variety of different ways to enjoy this mushroom: sautéing them or adding them to a stir-fry are two common, delicious ways. Use them in soups or stews for a meaty bite, or as an accent in a ramen bowl. They are also staples in risotto or, for some foodies, on white or red pizzas.

Shiitake is one of the most well-studied functional mushrooms. It's known to contain a compound called lentinan, which is highly supportive of the immune system. Shiitake is also widely used to support liver health. Packed with essential amino acids, protein, fiber, and B and D vitamins, Shiitake Mushrooms are a wonderful supplement for overall health. Like all functional mushrooms, Shiitake is considered an adaptogen, meaning it helps support the body's natural ability to deal with occasional stress and fatigue. Dried Shiitake, if left to dry outside in the sun, can potentially help you to meet the recommended daily dose of vitamin D. Like all functional mushrooms, it's important to dual extract your mushrooms with both alcohol and hot water to get both polar and nonpolar functional compounds for optimal health and wellness.

Shiitake As a Skin Irritant or Skin Healer

Some people have digestive issues when consuming certain mushrooms, Shiitake included. There are a few reports of people claiming they developed a rash from eating uncooked shiitake, so like all mushrooms, it's recommended you cook your mushrooms before eating them. Like most mushrooms, this mushroom is a bit paradoxical because it's traditionally used for supporting skin health. Since everyone is different, it's important to start with a small amount when trying anything new to see how it affects your system.

Snow Fungus

Tremella fuciformis

AT A GLANCE

GEOGRAPHIC LOCATION ✦ Global.

GROWING LOCATION ✦ Fungal parasite that attacks fungi in the *Annulohypoxylon* genus found on branches of broadleaf trees.

CHARACTERISTICS ✦ White, frond-like, and gelatinous.

PRIMARY USE ✦ Functional, edible, and for skin care.

Snow Fungus is also referred to as Silver Ear, White Jelly Mushroom, Snow Mushroom, *yín ěr*, *xuě ěr*, or *bái mù ěr* in Mandarin Chinese, *shiro kikurage* in Japanese, or *nấm tuyết* or *ngân nhĩ* in Vietnamese. This mushroom was first described in 1856 by one of the founders of plant pathology, British mycologist Miles Joseph Berkeley, from Brazilian collections found by British botanist Richard Spruce. The White Jelly Mushroom is now one of the most popular edible and functional species in the world.

Snow Fungus is a parasitic yeast that doesn't form fruiting bodies unless it parasitizes a fungus from the *Annulohypoxylon* genus. *Annulohypoxylon archeri*, or Black Fungus, is the fungus Snow Fungus prefers to grow on the most. Black Fungus can be found on branches of broadleaf trees, whether still attached or fallen. Therefore, you can find Snow Fungus in these locations as well.

This fungus is widely cultivated in China with supplemented hardwood sawdust bags containing both the Snow Fungus and Black Fungus cultures growing in unison. This type of growth is called the "dual culture method." Snow Fungus is cultivated in other Asian countries, but it's most popular in China, where hundreds of thousands of tons are grown every year.

The most popular uses of these mushrooms are for food, functional benefits, and skin care. For food, it's most commonly used to make a Chinese dessert soup called *luk mei* in Cantonese. In Vietnam, the mushroom is used to make *chè*, which is an umbrella term for any traditional pudding, sweet beverage, or dessert soup. Snow Fungus tastes fairly bland, but it has a unique gelatinous texture, making it good for soups.

Snow Fungus is widely used for skin care and has been traditionally used since 200 C.E. by royalty and the wealthy. Since this mushroom is a jelly fungus, it can hold up to five hundred times its weight in water, which is why it's used to nourish the skin. It's very helpful with moisturizing.

For functional purposes, Snow Fungus is packed with vitamin D, dietary fiber, and many other helpful compounds. These compounds help support your body's:

+ natural inflammatory response post workout
+ natural response to occasional stress
+ natural immune function
+ gut health
+ natural detoxification

Snow Fungus As Part of a Beauty Regimen

In ancient China, one of the "Four Great Beauties" was an imperial concubine named Yang Guifei. It was said that this woman used Snow Fungus regularly to maintain her youthful skin and glowing complexion.

Snow Fungus is also traditionally used for supporting healthy skin, bone health, cell and brain health, and more. It may not be possible to find dried or fresh Snow Fungus in your local grocery store if you live in North America. However, if you were looking to find it for functional or skin purposes, it's widely used in a lot of skincare and mushroom supplement products.

SNOW FUNGUS

SPLIT GILL

Split Gill

Schizophyllum commune

Split Gill is one of the most widespread mushrooms in the world. It's eaten in many countries, including China, Mexico, India, and many other tropical places. People in these warmer climates use the Split Gill in many recipes, such as chutney or salads, or roast them.

In China, the Split Gill is used as a functional mushroom and is commercially cultivated on a large scale. These cultivated mushrooms tend to be bigger than ones you find in the wild because they are given optimal environments to thrive in. That said, in the wild, they can dry out and rehydrate with the fluctuating temperatures, which gives them an evolutionary leg up. The other evolutionary advantage these mushrooms have is that they have a huge number of mating types. The Split Gill has 22,960 pairing possibilities. Since there are so many genetic variations available for this mushroom, it is able to adapt to pretty much every environment on the planet, including every continent except Antarctica. As with many mushrooms, some scientists question if all Split Gill species around the world are the same mushroom, or if further DNA analysis would break them apart into further subspecies.

The Split Gill Mushroom has some potentially unfortunate health effects for immunocompromised individuals. This mushroom may cause human mycosis (a disease that causes some loss of skin pigment when a fungus invades a person's tissues) in vulnerable individuals (especially children); if such an individual ingested a lot of spores, this fungus may grow inside them. The most extreme example of this was one case in a child where the fungi grew through the soft palate of her mouth and mushrooms actually fruited in her sinuses. However, this shouldn't be an issue for most people. Despite its negative effects on some people, this mushroom is often used as a functional mushroom. The Split Gill is said to have various compounds that help support the immune system.

These mushrooms are extremely resilient for other reasons too. Not only can they grow in humans, but another species of *Schizophyllum* was found underneath the Pacific Ocean. Researchers were drilling and taking up rock samples from the bottom of the ocean off the coast of Japan and found up to sixty-nine different types of fungi in the rock samples. One of the samples was a hibernating *Schizophyllum* species that they identified to be up to twenty million years old. When the scientists put the spores on a nutrient agar plate, not only did they germinate and form mycelium, but they also actually fruited mushrooms.

A Mushroom by One Name

Though it's likely to be split into multiple species eventually, the Split Gill Mushroom is one of few mushrooms that has kept its Latin name since it was first discovered. Elias Magnus Fries, Swedish mycologist, named this mushroom *Schizophyllum commune* in 1815, and it still has the same Latin name today.

Tarantula Cordyceps

Cordyceps caloceroides

The Tarantula Cordyceps is a spider-infecting mushroom that can be found in tropical regions in Mexico, Bolivia, Colombia, Brazil, Ecuador, and elsewhere. This is an unusual mushroom in that it infects a wide range of spiders in the *Mygalomophae* subclass including:

✦ bird-eating spiders
✦ trapdoor spiders
✦ tarantulas

Usually all *Cordyceps* are considered entomopathogenic fungi, meaning they parasitize insects, but the Tarantula Cordyceps is an interesting exception to the rule. Spiders are not classified as insects; they are arachnids. Therefore, this type of *Cordyceps* is technically classified as an araneopathogenic fungus, or a fungus that is pathogenic or parasitic to spiders.

To get to the spider, the Tarantula Cordyceps must lay in wait. Since this mushroom attacks spiders that dwell primarily on the ground, the spider needs to walk over the spores, or eat something containing the fungus, in order to get infected. The spiders of interest to the Tarantula Cordyceps Mushroom like to burrow deep in holes in the ground. For example, the trapdoor spider will make a secret trapdoor in the ground to hide its hole and surprise unsuspecting prey. If a Tarantula Cordyceps infects one of these trapdoor spiders, the mushroom then has the ability to push open that spider's trapdoor to spread its spores to the outside world. Regardless of its victim, the mushrooms that pop out of the spider must cover a large distance in order to break out of the soil. Some of the mushrooms can be as long as an adult's forearm; they can be deep red, orange, yellow, or white. Some scientists theorize that the mushroom will cover even a dead spider with mycelium, and the sticky substance then makes it easier to transfer spores to other things around it in the environment, even other spiders when they travel around and make contact with the spores.

Since this is an understudied mushroom, scientists are still unaware if it is found in other parts of the world. However, it's most likely distributed farther where other spiders of the *Mygalomorphae* subclass are located. Further research needs to be done on what compounds are present inside of this mushroom, and if it has any particular functional or scientific uses. A study of a closely related fungus that also grows on trapdoor spiders, *Cordyceps nidus*, showed antimicrobial content against *Staphylococcus* spp., *Enterococcus* spp., and *Bacillus* spp., among others. Future studies need to be done specifically with *Cordyceps caloceroides* to prove similar antimicrobial action.

Many Edible Mushrooms

Though the Tarantula Cordyceps is not classified as edible, many other mushrooms are. Close to twenty-two hundred species of mushrooms have now been classified as edible. That said, around two hundred of these species cause negative reactions in some people and/or are not safe to consume in their natural state (for example, they need to be cooked or otherwise treated).

TARANTULA CORDYCEPS

THE PRINCE

The Prince

Agaricus augustus

The Prince Mushroom is a choice edible species found growing in decaying organic matter in gardens or in deciduous and coniferous woods in Europe, North America, North Africa, and Asia. This mushroom was first described by Swedish mycologist Elias Magnus Fries in 1838 as *Agaricus augustus*. This is one of few mushrooms that has not had its Latin name changed since being discovered, at least not as of the writing of this book. For the name of this mushroom to not change for nearly two hundred years is rare, as Latin names of fungi are constantly changing as new species are found, a greater understanding of the scientific field is happening every day, and advancements with DNA technology are taking place. The last part of this mushroom's name, *augustus*, likely comes from the fact that it is commonly found in the month of August. However, *augustus* also translates to "noble or the illustrious one," leading to the mushroom's common name: The Prince.

Although this mushroom is a delicious edible species, it can be difficult to get a positive identification of it. Beginner mycologists and foragers should be very wary if they see what appears to be a Prince Mushroom in the wild, as it has many poisonous look-alikes. This mushroom can be confused with other poisonous *Agaricus* species, such as *Agaricus xanthodermus* (the Yellow-Staining Mushroom), *Agaricus moelleri* (the Inky Mushroom), and *Agaricus placomyces* (the Flat-Cap Agaric). Other toxic look-alikes could be poisonous *Amanita* species, including Death Cap Mushroom and Destroying Angel Mushroom species. So, although The Prince is highly sought after as a delicious edible, beginners should not forage for this mushroom, nor serve it to themselves or others, regardless of how certain they are they have found The Prince. Another thing to watch out for is that this mushroom tends to hyperaccumulate the metal cadmium. Experts should be careful if harvesting around factories or urban areas, which may have a lot of heavy metals in the soil.

If properly identified and harvested in a clean area by a knowledgeable forager, the Prince Mushroom is a choice find: It has a strong nutty, almondy, and anise smell and complements a lot of dishes. This mushroom can be used in the same way one might use a Portobello Mushroom. The best way to preserve it is by drying it and then, when ready to cook, rehydrating it in milk or water. The Prince Mushroom can be added to pizzas, tapas dishes, soups, pastries, sauces, stews, or even stuffed and baked. They are also delicious simply sautéed or grilled.

> **Other Fun Facts about The Prince**
>
> The Prince actually starts growing in the early summer, and depending on where you live, you may be able to find this mushroom through late October. Another identifying characteristic is that the spore print is usually a warm, dark-chocolatey brown.

Tiger Sawgill

Lentinus tigrinus

AT A GLANCE

GEOGRAPHIC LOCATION ✦ Europe, North America, Asia, and Australia.

GROWING LOCATION ✦ A saprophytic mushroom, it grows on decaying wood in frequently flooded areas near rivers or other bodies of water.

CHARACTERISTICS ✦ A cap-and-stem mushroom with a dimple on the cap. This mushroom is whitish tan with darker brown flakes on the cap and white gills underneath.

PRIMARY USE ✦ Edible, functional, and mycoremediation.

Tiger Sawgill, also called Tiger Agaric, is an edible, functional mushroom that also has potential applications in mycoremediation. It was originally named *Agaricus tigrinus* in 1782 by French botanist Jean Baptiste François Pierre Bulliard; Bulliard also gave it the common name *L'Agaric Tigré*. It wasn't until 1825 that the name was officially changed to its current Latin name, *Lentinus tigrinus*, by Swedish mycologist Elias Magnus Fries. *Lentinus* comes from Latin, roughly translating to "resembling pliable," probably referring to the flexibility of the mushroom flesh. *Tigrinus* means "like a tiger," referring to the dark brown scales on the top of the cap that can sometimes resemble tiger stripes.

Tiger Sawgill is found in Europe, North America, Asia, and Australia. This mushroom is saprophytic, meaning that it takes nutrients and grows on old rotten trees. The Tiger Sawgill tends to like trees near riverbanks that get frequent overflooding of river water, specifically elm, willow, cottonwood, and silver maple.

A 2004 study of this mushroom showed a clear difference in DNA between the North American variety of Tiger Sawgill and the version that appears elsewhere. Although these two varieties of Tiger Sawgill can still mate and transfer genetic information on a petri dish in a lab, one of these species is likely to get a new Latin name.

Two famous "cousins" of *Lentinus tigrinus* are *Lentinula edodes*, commonly known as Shiitake, and *Neolentinus lepideus*, commonly known as the Train Wrecker. Shiitake (see earlier entry in this book) is one of the most widely eaten gourmet mushrooms in the world and is also a highly revered functional mushroom for supporting the immune system and liver health. *Neolentinus lepideus* is also called the Train Wrecker because it's found on railroad ties, degrading them over time, which results in a lot of destruction. The Train Wrecker Mushroom was apparently important in the winning of the American Civil War because it grew on railroad ties in the Confederate South and disrupted the Southerners' routes to get crucial supplies. This disruption gave the Union an upper hand.

Although Tiger Sawgill is an edible mushroom, it's not as revered as its cousin, Shiitake. The Tiger Sawgill is also a functional mushroom and contains compounds which offer antioxidant support and immune system support.

The Tiger Sawgill is also studied for mycoremediation. It may be important for environmental remediation of biological contaminants, especially in dirty waterways. This mushroom uses its mycelial web to trap contaminants, cleaning the water.

Toadstool versus Mushroom

People may think that toadstools are different from mushrooms. However, that's not the case. Toadstools are typically thought to be colorful and inedible—like the bright red variety with spots. At the end of the day, they're synonymous, with toadstools just being a fun name for mushrooms.

TIGER SAWGILL

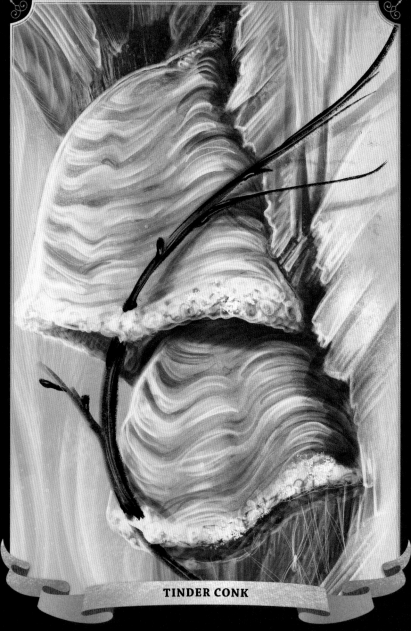

TINDER CONK

Tinder Conk

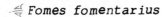
Fomes fomentarius

The Tinder Conk, also called the Tinder Fungus, Hoof Fungus, Tinder Polypore, or Ice Man Fungus, is a hoof-shaped conk mushroom that grows on the side of various trees in North America, Asia, Africa, and Europe. On the east coast of the United States, you can usually find them on birch trees, but they also grow on beech, oak, maple, cherry, hickory, lime, poplar, willow, alder, hornbeam, sycamore, and softwoods too.

The Tinder Conk has a long history of name changes:

+ 1753: First called *Boletus fomentarius* by Swedish botanist Carl Linnaeus.
+ 1783: Changed to *Agaricus fomentarius* by French naturalist Jean-Baptiste Lamarck.
+ 1818: Renamed *Polyporus fomentarius* in 1818 by German mycologist Georg Friedrich Wilhelm Meyer.
+ 1849: Finally, Swedish mycologist Elias Magnus Fries solidified the name as *Fomes fomentarius*.

As for the meaning of the name, *fomentarius* comes from the Latin *fomentum*, meaning "tinder." Its name refers to the conk's ability to be used as tinder for fire making. This mushroom when dried can easily have a hole drilled in it where an ember of a fire can be placed. It's easy to carry during travels in order to start fires. The Tinder Conk Mushroom has an incredible ability to slowly smolder without going out, making it useful for carrying around embers. Or if dried, sliced thin, shredded, and fluffed, this makes a great fire starter.

Ötzi the Iceman, the five-thousand-year naturally preserved mummy who was found in the Alps, had *Fomes fomentarius* in his pack. He was also found with tattoos on his body that correlate with meridian lines and acupuncture spots. Scientists speculate that he may have used the smoldering Tinder Conk under these areas to bring heat and stimulate chi or energy flow, to help with his osteochondrosis and spondylosis (disorders affecting the bones and cartilage). This process, called moxibustion, is usually done with burning mugwort, but this mushroom could have been used as a substitute.

The Tinder Conk is a functional mushroom. It is used to:

+ support immune function
+ support the body's natural inflammatory response post workout
+ offer antioxidant support
+ support healthy blood sugar levels

Elias Magnus Fries

Elias Magnus Fries (1794–1878) was one of the best-known mycologists of all time. He is considered the founder of fungal taxonomy, and wrote a great deal about mushrooms, giving many their current scientific names, even though he was born over two hundred years ago. He was from Sweden, and went to school at Lund University, getting a doctorate in 1814.

Tropical Cinnabar Bracket Fungus

Pycnoporus sanguineus

The Tropical Cinnabar Bracket Fungus will stop you dead in your tracks if you're exploring a tropical or subtropical jungle. This mushroom's bright reddish-orange color is unbelievably eye-catching, even from quite a distance. This bright color is pretty common in all *Pycnoporus* species of mushrooms, such as Cinnabar Polypore (*Pycnoporus cinnabarinus*), the Southern Cinnabar Polypore (*Pycnoporus coccineus*), and *Pycnoporus puniceus*. Interestingly, though these mushrooms are all very similar, you can drop 5 percent of a particular chemical solution on the cap of all four of these mushrooms and they will all turn black, with the exception of the Tropical Cinnabar Bracket Fungus, which will turn a greenish brown when the mushroom dries. First discovered either in Guam or Guana Island, this mushroom is extremely common in tropical regions all over the world.

As a functional mushroom, the Tropical Cinnabar Bracket Fungus is widely used throughout Asia and elsewhere:

✦ In traditional Chinese herbalism, it is used in a tea for cooling and detox, as well as supporting the body's normal inflammatory response and immune system. This mushroom is said to invigorate the body's vital energy and also be supportive for the cardiovascular system, as well as for skin health.

+ Traditionally in Malaysia, Tropical Cinnabar Bracket Fungus is used for skin health as well, as the red color is thought to ease naturally occurring inflammation in the skin. In Java, an island in Indonesia, the mushroom is extracted in warm oil and used on the skin as well.

+ A pigment called cinnabarin is extracted from Tropical Cinnabar Bracket Fungus and used in the textile industry for the decolorization (or removal of color from something) of certain dyes.

+ In the biotechnology space, some researchers are looking to use the Tropical Cinnabar Bracket Fungus to absorb heavy metals within the bloodstream. This is not the only space where this mushroom is being researched for its effects on toxins. Environmentally, this mushroom is also used in the bioremediation of crude oils. In the Sucumbíos region of Ecuador, a log was completely submerged in an unlined oil pit and Tropical Cinnabar Bracket Fungus was found growing on that log. The growth of the mushroom seemed astounding, and it was taken into the lab by researchers from the grassroots bioremediation group CoRenewal. These researchers ran tests and found evidence showing that the Tropical Cinnabar Bracket Fungus was effective at remediating oil on petri dishes.

Other Health-Related Uses

The Tropical Cinnabar Bracket Fungus has a great number of uses that are still being discovered. Some of them are shown to help with symptoms of diseases like sore throats, ulcers, arthritis, gout, toothaches, bleeding, and fevers. It can also help protect against E. coli, thanks to its antibacterial nature.

TROPICAL CINNABAR BRACKET FUNGUS

TURKEY TAIL

Turkey Tail

Trametes versicolor

AT A GLANCE

GEOGRAPHIC LOCATION ✦ Global.

GROWING LOCATION ✦ Decaying wood.

CHARACTERISTICS ✦ Fan-shaped mushroom with multicolored bands ranging from blue to white to purple to black to brown to gray.

PRIMARY USE ✦ Functional, mycoremediation, and dyeing

Turkey Tail is also called *Kawaratake* in Japan, meaning "mushroom by the riverbank," and *Yun-Zhi* in China, meaning "cloud fungus." Its Latin species name, *versicolor*, meaning various colors, was given because this mushroom displays up to a half a dozen different colored bands in a fan shape that looks almost like the tail of a turkey. You can find Turkey Tail Mushrooms growing on dead logs in the woods. This mushroom often grows in clusters, thus looking even more like a turkey's plumage. The top of the Turkey Tail's cap is subtly hairy, while the bottom is pure white. The bottom of the cap shows the mushroom's miniscule pores, where the spores shoot out of.

There are a few look-alikes to this mushroom, including the False Turkey Tail (*Stereum ostrea*), Crowded Parchment Fungus (*Stereum complicatum*), Violet-Toothed Polypore (*Trichaptum biforme*), Hairy Bracket (*Trametes hirsuta*), and others. One of the most telltale signs for identifying this mushroom is the spore surface at the bottom. The porous surface should be pure white and filled with tiny pores (at least three pores per millimeter).

Turkey Tail is important within the natural world. Its role in the ecosystem as a white rot fungus means that the mushroom helps break down wood to aid in nutrient cycling for the soil and the surrounding plants. As this nutrient cycle happens, more nutrients are released into the soil, thus helping the other plants in the area.

Turkey Tail for Mycoremediation

Turkey Tail Mushrooms have such a wide variety of uses. As the science behind the process improves, more species of mushrooms are being used for mycoremediation—the Turkey Tail is one such species. It's used in commercial mycoremediation for cleaning various biological contaminants, chemicals, and heavy metals.

The Turkey Tail Mushroom is wildly commercially cultivated, using logs outside or hardwood sawdust bags indoors, for the functional supplement market. In traditional Chinese herbalism, this mushroom is used for supporting energy as well as spleen and heart health. The Turkey Tail Mushroom is also used for supporting the immune system, respiratory system, and urinary system. Packed with two useful compounds, this mushroom is most commonly used to support the immune system. Also, being an adaptogen, meaning it helps the body adapt to occasional stress, Turkey Tail is helpful for supporting the body's ability to deal with occasional stress and fatigue. It's best to utilize this mushroom by doing a dual extraction, meaning using alcohol and hot water to access all the polar and nonpolar functional compounds to support overall health and wellness.

Although it does not impart any functional benefits for health and wellness, some people use Turkey Tail like a chewing gum. Some claim to thoroughly enjoy replacing their store-bought gum with this organic substitute. The last potential use that is currently known for Turkey Tail is to use it for dyeing wool, paper, and other fabrics by adding ammonia to some of the mushroom. You then get a brown color that you can use for dyeing.

Umbrella Polypore

Polyporus umbellatus

AT A GLANCE

GEOGRAPHIC LOCATION ✦ North America, Europe, and Asia.

GROWING LOCATION ✦ Primarily on the roots of old beech or oak trees.

CHARACTERISTICS ✦ These mushrooms often grow in clusters. They are whitish-gray to yellow to brown cap-and-stem umbrella-shaped mushrooms with pores.

PRIMARY USE ✦ Edible and functional.

The Umbrella Polypore is known as *Zhu Ling* ("hog tuber") in China and *Chorei Maitake* ("wild boar's dung maitake") or *Tsuchi Maitake* ("earth maitake") in Japan. It gets these names because it grows from a black sclerotium (an underground mass of mycelium), also called a tuber, that looks like a boar's droppings. Maitake is included in some of the common names because it looks almost like a Maitake Mushroom. The Latin name *Polyporus umbellatus* means "many-pored umbrellas," relating to the caps that look like umbrellas with pores on the bottom.

Finding a spot where the Umbrella Polypore grows is very rare, and in some parts of Europe the spots are guarded and passed down in the family. The sclerotium, being very large, will stay in the ground year after year, and mushrooms will pop up from the same spot in the ground annually. Some expert foragers say that they have gone over forty years looking for this mushroom and have never found it.

In China, Umbrella Polypore is cultivated and used extensively in traditional Chinese herbalism. Chinese herbalists use the tuber to support the immune system, help water flow in the body, help the body with its natural inflammatory response post workout, and offer antioxidant support.

The Umbrella Polypore is an edible mushroom that tastes woodsy, with a distinctive, traditional mushroom taste. This taste gets stronger as the mushroom ages, so if you have a delicate palate, it's best to find them when they are young. One way to cook this mushroom that is popular for *Polyporus umbellatus*, Maitake, and Oyster Mushrooms is to use them to make mushroom "steaks." To do this, take a cluster of the mushrooms, put them on a nonstick or cast iron pan, turn the heat on medium high, and then put a weight on top to smush it down. This weight can easily be created with another heavy cast iron pan. Some people might also add a cooking sauce that the mushrooms will soak up to enhance the flavor. However, with something as rare as the Umbrella Polypore, it might be best to keep it as plain as possible to preserve the natural flavor. Once the "steak" is golden brown, flip it. This dish makes a perfect vegetarian main course.

A Rare Piece of Mushroom Art

Some people adore the Umbrella Polypore so much that they cannot bear the thought of eating it; they prefer it as art. Considering how rare it is to find, some have dried this mushroom, coated it in an epoxy resin to preserve it, and put it on their altar or mantel for decoration.

UMBRELLA POLYPORE

VERDIGRIS AGARIC

Verdigris Agaric

Psilocybe aeruginosa or *Stropharia aeruginosa*

The Verdigris Agaric is a stunning bright blue cap-and-stem mushroom. It typically grows on wood chips in garden parks all over Europe, Asia, and North America. British botanist William Curtis first described this species as *Agaricus aeruginosus* in 1782. The mushroom's name was later changed to *Stropharia aeruginosus* by French mycologist Lucien Quélet in 1872. The name was again changed, with much controversy, to *Psilocybe aeruginosa* in 1995 by Dutch mycologist Machiel Noordeloos. This name change was controversial because there are many reports that this mushroom is psychoactive, containing psilocybin and psilocin, which are the two main molecules in Magic Mushrooms (see earlier entry in this book). However, American mycologists Michael W. Beug and Jeremy Bigwood published a paper in 1981, called "Psilocybin and Psilocin Levels in Twenty Species from Seven Genera of Wild Mushrooms in The Pacific Northwest, U.S.A.," in which they revealed that samples of *Stropharia aeruginosa* that were tested in 1979 contained zero traces of either psilocybin or psilocin. There is one other *Psilocybe* species, called *Psilocybe fuscofulva*, that does not contain either psilocybin or psilocin, but this is another highly debated species that mycologists are petitioning to kick out of the genus *Psilocybe* because of the lack of these two compounds. Most guidebooks still refer to Verdigris Agaric as *Stropharia aeruginosa* instead of *Psilocybe aeruginosa*; at this point the name change is still controversial.

For added controversy, many guidebooks in the United States state that this mushroom is poisonous, but guidebooks in Europe say that this species is edible but a little spicy. Such confusion and debate on whether this mushroom is edible or not, toxic or not, and psychoactive or not could be due to multiple species being included under the same Latin name. This happens all the time in mycology. Multiple species having the same Latin names will continue to change as we get better results from DNA analysis. There could be a psychoactive version of the Verdigris Agaric out there that contains psilocybin and psilocin, and another nonpsychoactive version that doesn't, and the same goes for edibility and toxicity. The Verdigris Agaric is something that deserves more research as it's a truly beautiful mushroom.

Some studies show this mushroom can be used for functional benefits such as supporting the immune system and brain function. As stated above, because of its uncertainty, ingesting this mushroom is not advised until more research is done.

The Verdigris Agaric Dulls As It Ages

Though these mushrooms are beautiful and bright blue when they are young, they often lose their lustrous color as they age. As they get older, the color in the mushroom's gluten fades, and the center becomes yellowish to a warm tan in color. If it lives long enough, the mushroom will lose all its blue color entirely.

Violet-Toothed Polypore

⋚ *Trichaptum biforme*

The Violet-Toothed Polypore Mushroom is one of the most useful to know when foraging in a hardwood forest in North America, as it pops up pretty much everywhere from May to December. This mushroom is very small and unassuming at first, but once you turn it over or peek underneath, it may stun you with its unique violet color. The color is stronger in the younger specimens and fades with age.

The Violet-Toothed Polypore sometimes has an accompanying blanket of moss on the top. If you squint while looking at the moss, you might see some tiny black club-like mushrooms called *Phaeocalicium polyporaeum*; these are saprobic ascomycetes (a type of fungus that eats organic matter) that like to grow on the upper surface of older Violet-Toothed Polypore Mushrooms.

In the wild, the Violet-Toothed Polypore is not the easiest for new mushroom enthusiasts to identify. This mushroom is easily confused with Turkey Tail Mushrooms (see earlier entry in this book), as they may be similar in color; also, there are many Turkey Tails look-alikes such as *Stereum ostrea*, *Stereum complicatum*, and others. However, the Violet-Toothed Polypore can be distinguished from its look-alikes by its distinct, lilac-colored band on the top and the violet-colored underside. The colors on this species are much brighter than on a Turkey Tail Mushroom. However, there is a look-alike that looks nearly identical to the Violet-Toothed Polypore; this is *Trichaptum abietinum*. The only distinguishable difference between these two mushrooms is that *T. abietinum* grows on conifers. That said, both mushrooms are inedible but are not poisonous.

Traditionally this mushroom was used by the NLaka'pamux First Nations people of British Columbia to give strength to young men. They called it *Kalulaa'iuk*, or "owl wood," and prepared it by collecting its spores and using it as a rub. Modern-day medicine is still researching to learn the potential health benefits of this mushroom. However, it does have one functional purpose: to support immune health.

One indicator that a fungus might be good for mycoremediation is if it's seen growing everywhere and in challenging environments. When these indicators are met, it means the mushroom has the enzymes and hardiness to fight off competition and survive. The Violet-Toothed Polypore's popularity in the forest indicates that this is a great candidate. This mushroom is used for remediation of PCPs and scientists are also looking to see if this could be a good candidate to use for remediation of toxic chemicals in colder climates.

Violet-Toothed Polypore As a Natural Dye

Organic dyes are surprisingly potent, and definitely a great alternative to use for dyeing paper or yarn. The beautiful purple color of the Violet-Toothed Polypore can be extracted and used for this purpose.

VIOLET-TOOTHED POLYPORE

WHITE BASKET FUNGUS

White Basket Fungus

 Ileodictyon cibarium

The White Basket Fungus is one of the oddest looking and weirdest smelling mushrooms in the fungal kingdom. This mushroom looks like a white, geometric cage-like structure, and it's covered in a brown slime. If that hasn't frightened you off yet, the smell might. The White Basket Fungus smells absolutely foul. The smell is part of its reproduction cycle. The White Basket Fungus's brown slime is encased in spores, and the mushroom uses the foul-smelling odor to attract flies to the slime. The flies then disperse the mushroom's spores once they leave. The smell means this mushroom is best appreciated from a distance and with your nose plugged. The shape of the mushroom's "cage" has been studied by scientists because its geometric shape is thought to be extremely stable, given its strength-to-weight ratio.

In order to help spread its spores, the White Basket Fungus actually mimics a tumbleweed. It separates itself from the structure connecting it to the ground, and then the wind rolls the round mushroom husk. This way, the mushroom attracts even more flies.

This mushroom has been used traditionally in New Zealand as a food source. The Māori people eat the egg-like structure of the mushroom before it develops a foul odor and expands into its geometric shape. Within the "egg," there is a tan outer layer, an inner white layer, and a dark-brown center. The Māori have over thirty-five different names for this mushroom including:

- ✦ *Tutae Kehua*, meaning "ghost droppings"
- ✦ *Whareatua*, or "house of the devil"
- ✦ *Tūtae Whetū*, translated as "star dung"

Some other names refer to thunderstorms, as it often appears after rainfall. It's inadvisable to eat the White Basket Fungus when it's fully mature and starting to smell. But before the mushroom gets to that point, some Māori like to cook the "eggs" in a hāngī, which is a pit underground that is traditionally used to steam food with hot rocks or coals.

This mushroom was almost picked as New Zealand's national fungus but came in second behind the Blue Pinkgill (see earlier entry in this book) in 2018.

To preserve a dry specimen, people sometimes blow up a balloon inside the White Basket Fungus, tie off the balloon, let the mushroom dry, and then pop the balloon. What's left behind is a dried but intact cage structure of the mushroom for display. As one of the coolest-looking mushrooms in the world, it would definitely make a beautiful art piece, or a mycology trophy of sorts.

How to Get Rid of White Basket Fungus

If you are not interested in the foul-smelling fungus that cropped up in your garden, you will have to do some upkeep. You will need to remove any mushrooms, hopefully at the egg-like stage, as you do not want the spores to spread. Be on the lookout for other developing mushrooms throughout the year. You might want to think twice about getting rid of them, though, because even though these mushrooms smell terrible, they help degrade organic matter in your garden and give your plants more nutrients.

Wine Cap

Stropharia rugosoannulata

Wine Cap, also called King Stropharia or Garden Giant, is one of the easiest mushrooms to grow and is considered a choice edible. In 1922, American mycologist William Alphonso Murrill, drawing upon an earlier description by American botanist William Gilson Farlow, named this mushroom *Stropharia rugosoannulata*. This mushroom's name remains the same today. *Stropharia* comes from the Greek *stophos*, meaning "melt." *Rugosoannulata* comes from the Greek words *rugoso*, meaning "wrinkled," and *annulata*, meaning "a ring."

One of the key features of the Wine Cap is its notable ring, which is left by the parted veil around the stem of the mushroom; this ring looks like a wrinkly cogwheel. This mushroom is extremely resilient and can thrive in both direct sunlight and shade. The amount of sunlight the mushroom gets will affect the cap color: It ranges from deep burgundy in the shade to a very washed-out light brown in direct sunlight. The color variation on the cap sometimes makes identification difficult; you can't always rely on the burgundy color.

Another way to identify the Wine Cap Mushroom is its incredibly thick and ropy mycelium. When you pull up the mushroom, the mycelium carries along with it a string of wood chips, showing impressive tensile strength. You can also cut off the "butt" of the mushroom with attached mycelium and replant it into wood chips and the mycelium will continue growing.

There are many ways to grow Wine Caps. Some people like to collect the mushroom butts and throw them into a blender with water and molasses—to enhance growth—to create a slurry which they then pour onto wood chips to continue the growth of this mushroom. Another easy technique is to buy a sawdust block online from a reputable vendor, mix it with fresh hardwood chips, let it grow for six months or so, and then use that new pile of wood chips as spawn for more wood chip piles. This process can be repeated for as long as you have a supply of fresh hardwood chips to keep feeding the Wine Caps.

Wine Caps make great garden companions underneath the shade of any plants. In Europe, people grow the mushrooms under corn. Wine Caps have been known to attack nematodes by deploying spiky cells (also called spur cells), which immobilize and eat the nematodes in your garden. Research has shown that the mycelium of the Wine Cap Mushroom can be used to remediate *E. coli* in waterways.

Delicious Wine Caps

Wine Caps are a delicious edible mushroom. The most common way to cook them is by sautéing them in oil or butter with salt and pepper. If you want something a little more challenging, bake the caps in the oven with tomatoes, cheese, basil, balsamic vinegar, and eggplant.

WINE CAP

WITCHES' BUTTER

Witches' Butter

⤳ *Tremella mesenterica*

AT A GLANCE

GEOGRAPHIC LOCATION ✦ Temperate and tropical regions of Africa, Asia, Australia, Europe, and North and South America.

GROWING LOCATION ✦ On dead branches as a parasite of fungi in the genus *Peniophora*.

CHARACTERISTICS ✦ Yellow, gelatinous, and intestine shaped.

PRIMARY USE ✦ Edible and functional.

Witches' Butter, also called the Yellow Brain Mushroom or the Golden Jelly Fungus, is a parasitic mushroom that attacks fungi in the genus *Peniophora*. Witches' Butter grows in temperate and tropical regions of Africa, Asia, Australia, Europe, and North and South America on dead branches. It usually appears on thin branches of hardwoods. This mushroom has an incredibly short shelf life, only showing its true gelatinous form for a few days after the rain falls. The mushroom then shrivels up and dries into a thin film, only to pop back up after it rains again.

Witches' Butter was originally called *Helvella mesenterica* in 1774 by German mycologist Jacob Christian Schäffer. The mushroom got its current Latin name, *Tremella mesenterica*, in 1822 from Swedish mycologist Elias Magnus Fries based on an earlier description by Swedish botanist Anders Jahan Retzius in 1769.

Tremella comes from the Latin word *tremere*, meaning "to tremble." *Mesenterica* is derived from ancient Greek, meaning "middle intestine," which refers to the gelatinous intestine shape of the fungus. This mushroom's common name has an interesting origin. Witches'

Butter's name is based on Swedish folklore. In folklore, it's said that the devil gives witches beasts, called carriers, and the yellow bile that the carriers threw up was called the "butter of the witches."

A common look-alike for this mushroom is the yellow *Tremella aurantia*, which is a parasitic fungus that grows on *Stereum hirsutum* (the False Turkey Tail Fungus). Another look-alike is *Dacrymyces chrysospermus* (the Yellow-Orange Jelly Fungus), which also has a similar color and texture.

Although Witches' Butter is not highly sought after for food and is described as edible but flavorless in many field guides, it's actually sought after in China. There is a famous Chinese soup made with Witches' Butter, lotus seeds, lily bulbs, jujube, and other ingredients. This soup is supposed to be functional in nature: It is said to be supportive to the immune system and cooling.

Witches' Butter does appear to have some other health benefits. It contains a compound that helps support the immune system, healthy blood sugar levels, normal inflammatory response after a workout, cardiovascular health, and the body's response to seasonal changes.

The Unique Umami Taste of Mushrooms

Mushrooms have one of the hardest tastes to come by: umami. Umami has a savory taste and is often linked to cooked meats. Mushrooms are one of the few nonanimal life-forms that has this taste. Raw mushrooms themselves do not have as much umami flavor as cooked mushrooms. Cooking with this flavor adds an unexpected, delicious twist. Witches' Butter is not a mushroom with a strong umami taste.

Wood Ear Mushrooms

◄ *Auricularia* spp.

The Wood Ear Mushroom species get their name from both their appearance—as they look like a human ear—and the way the flesh feels—like a human ear if you rub it between your fingers. These mushrooms used to fall under the Latin name *Auricularia auricula* or *Auricularia auricula-judae*, but recent DNA analysis has split this mushroom and its Latin name into many smaller species. These include *Auricularia americana*, *Auricularia fuscosuccinea*, and *Auricularia angiospermarum* in the United States; *Auricularia auricula-judae* in Europe; and *Auricularia heimuer* and *Auricularia cornea* in China.

Another common name for this species is Judas's Ear. It got that nickname after the biblical story about Judas Iscariot. The biblical story says that Judas betrayed Jesus, then hanged himself. These mushrooms are said to be Judas's remains.

The most common cultivated Wood Ear varieties in China (*Auricularia cornea*, called *Maomuer*, meaning "cloud ear," and *Auricularia heimuer*, called *Heimuer*, meaning "black wood ear") are used in various dishes and in functional ingredients. *Auricularia cornea* was originally described in Hawaii by German naturalist Christian Gottfried Ehrenberg in 1820.

These mushrooms are typically picked and cultivated because they are a culinary treat. In China, billions of kilograms of Wood Ear Mushrooms are grown annually; it is one of the most commonly cultivated mushrooms in the world. Like tofu, these mushrooms absorb the flavor of whatever is in the dish, but they will add a slight earthy flavor. Wood Ears can easily be prepped for salads, soups, and even teas. They are also good in stir-fries.

Wood Ear Mushrooms are also used for their functional benefits. *Auricularia polytricha* is known for many helpful compounds for the human body. *Auricularia cornea* is used in traditional Chinese herbalism for supporting the immune system and heart function, for mental and physical energy, for giving lightness and strength to the body, and for strengthening the will. It's also used to support the lungs, circulatory system, stomach health, throat, and eye health.

As with all mushrooms and other functional foods, caution should be exercised when using them. For example, there is a story that a researcher accidentally cut himself on a broken piece of glassware and was not able to control the bleeding. No one could figure out why and he was asked about what he had eaten recently. He said he'd been eating a lot of sweet and sour soup while he had been stuck in the lab. The scientists deduced that the *Auricularia cornea* in the sweet and sour soup had anticoagulant properties that had thinned his blood.

WOOD EAR MUSHROOMS

ZOMBIE-ANT FUNGUS

Zombie-Ant Fungus

 Ophiocordyceps unilateralis

The Zombie-Ant Fungus was first discovered by British naturalist Alfred Russel Wallace in 1859. This type of fungus is known as an entomopathogenic fungus, meaning the fungus attacks insects, primarily ants in this case. Here is how the fungus spreads: Ants of the tribe *Camponotini* become infected in various ways: a spore lands on an ant, an ant eats something with the fungus inside of it, or an ant walks over a spore and the spore sticks to its body. When contact is detected, the spore begins to germinate and releases certain enzymes to breach the exoskeleton of the ant. Once the fungus is inside, it turns into a yeast-like growth. It grows rapidly inside the body of the insect. The fungus will also infect the ant with nerve toxins to influence the ant's behavior and wrap its mycelium around the muscles of the ant to influence its movement. Usually by about ten days after infection, the fungus controls the ant's muscles and uses its nerve toxins to make the ant climb up a tall tree or plant. The fungus-controlled ant will find a spot with the right humidity level and at the right angle to maximize spore release into the wind. The cycle continues as the fungus spreads its spores onto unassuming ants on the forest floor below.

Zombie-Ant Fungus's Death Grip

Right before the fungus fruits, the zombie-ant that it's controlling will bite down onto the vein of a leaf and be unable to let go. This bite is known as the death grip. Oddly enough, most likely the ant will be upside down on a leaf when the death grip happens. Four to ten days later, a mushroom will pop out of the zombie-ant's head, raining spores down on the forest floor below.

The Zombie-Ant Fungus has been known to destroy entire ant colonies, so usually when you find one infected ant, you're more than likely to find others in the nearby area. Sometimes, if you're lucky, you can find dozens or hundreds of infested ants on a single tree. This phenomenon is known as an ant graveyard. If you're looking for these fungi in the wild, they can be easy to spot: Scan at eye level, underneath leaves, and look for any black spots.

There is some serious medical potential for the Zombie-Ant Fungus. The fungus contains known metabolites, as well as several untapped compounds that are reportedly being investigated for antitumor, immunomodulatory, and anticancer properties. Some of the pigments that can be made from this fungus are also being studied for use as dyes for food and cosmetics and in pharmaceuticals.

ABOUT THE AUTHOR

Alex Dorr is the founder and CEO of the Austin-based functional mushroom company Mushroom Revival Inc. Alex hosts the number one mushroom podcast in the world, *Mushroom Revival Podcast*, and authored the book *Mycoremediation Handbook: A Grassroots Guide to Growing Mushrooms and Cleaning Up Toxic Waste with Fungi* (2017). He was recently nominated as one of Austin Inno's 25 Under 25, and is absolutely obsessed with the power of mushrooms.

ABOUT THE ARTIST

Sara Richard is an Eisner and Ringo Award–nominated artist from New Hampshire. Her art is inspired by Art Nouveau, Art Deco, funerary imagery, and the natural world. Her creations tend to skew into the macabre and unknown with a balance of sweetness and sentimentality, honoring the Victorian-era theme of *memento mori*. As a native of New Hampshire, Sara grew up surrounded by trees and plenty of wild mushrooms. When not making art or writing, she's watching horror movies, cleaning forgotten gravestones with her mom, and collecting possibly haunted curios from the nineteenth century. Her online gallery can be found at SaraRichard.com.